cradle to cradle

Remaking the Way
We Make Things

cradle to cradle

William McDonough & Michael Braungart

North Point Press
A division of Farrar, Straus and Giroux
New York

North Point Press
A division of Farrar, Straus and Giroux
19 Union Square West, New York 10003

Distributed in Canada by Douglas & McIntyre Ltd.
Printed in China
First edition, 2002

DuraBook™, patent no. 6,773,034, is a trademark of Melcher Media, Inc.,
124 West 13th Street, New York, NY 10011, www.melcher.com.
The DuraBook™ format utilizes revolutionary technology
and is completely waterproof and highly durable.

Library of Congress Cataloging-in-Publication Data
McDonough, William.
Cradle to cradle: remaking the way we make things / William McDonough
and Michael Braungart.
p. cm.
ISBN-13: 978-0-86547-587-8
ISBN-10: 0-86547-587-3 (hc. : alk. paper)
1. Recycling (Waste, etc.) 2. Industrial management—Environmental
aspects. I. Braungart, Michael, 1958– II. Title.

TD794.5 .M395 2002
745.2—dc21

2001044245

Designed by Janine James / The Moderns

www.fsgbooks.com

15 17 16

To our families,
and to all of the children of
all species for all time

The world will not evolve past its current state of crisis by using the same thinking that created the situation.

—ALBERT EINSTEIN

Glance at the sun.
See the moon and the stars.
Gaze at the beauty of earth's greenings.
Now, think.

—HILDEGARD VON BINGEN

What you people call your natural resources our people call our relatives.

—OREN LYONS, *faith keeper of the Onondaga*

Contents

cradle to cradle

Introduction

This Book Is Not a Tree

At last. You have finally found the time to sink into your favorite armchair, relax, and pick up a book. Your daughter uses a computer in the next room while the baby crawls on the carpet and plays with a pile of colorful plastic toys. It certainly feels, at this moment, as if all is well. Could there be a more compelling picture of peace, comfort, and safety?

Let's take a closer look. First, that comfortable chair you are sitting on. Did you know that the fabric contains mutagenic materials, heavy metals, dangerous chemicals, and dyes that are often labeled hazardous by regulators—except when they are presented and sold to a customer? As you shift in your seat, particles of the fabric abrade and are taken up by your nose, mouth, and lungs, hazardous materials and all. Were they on the menu when you ordered that chair?

That computer your child is using—did you know that it contains more than a thousand different kinds of materials, including toxic gases, toxic metals (such as cadmium, lead, and mercury), acids, plastics, chlorinated and brominated substances, and other additives? The dust from some printer toner cartridges has been found to contain nickel, cobalt, and mercury, substances harmful to humans that your child may be inhaling as you read. Is this sensible? Is it necessary? Obviously, some of those thousand materials are essential to the functioning of the computer itself. What will happen to them when your family outgrows the computer in a few years? You will have little choice but to dispose of it, and both its valuable and its hazardous materials will be thrown "away." You wanted to use a computer, but somehow you have unwittingly become party to a process of waste and destruction.

But wait a minute—you care about the environment. In fact, when you went shopping for a carpet recently, you deliberately chose one made from recycled polyester soda bottles. Recycled? Perhaps it would be more accurate to say *downcycled*. Good intentions aside, your rug is made of things that were never designed with this further use in mind, and wrestling them into this form has required as much energy—and generated as much waste—as producing a new carpet. And all that effort has only succeeded in postponing the usual fate of products by a life cycle or two. The rug is still on its way to a landfill; it's just stopping off in your house en route. Moreover, the recycling process may have introduced even more harmful additives than a conventional product contains, and it might be off-gassing and abrading them into your home at an even higher rate.

The shoes you've kicked off on that carpet look innocuous enough. But chances are, they were manufactured in a developing country where occupational health standards—regulations that determine how much workers can be exposed to certain chemicals—are probably less stringent than in Western Europe or the United States, perhaps even nonexistent. The workers who made them wear masks that provide insufficient protection against the dangerous fumes. How did you end up bringing home social inequity and feelings of guilt when all you wanted was new footwear?

That plastic rattle the baby is playing with—should she be putting it in her mouth? If it's made of PVC plastic, there's a good chance it contains phthalates, known to cause liver cancer in animals (and suspected to cause endocrine disruption), along with toxic dyes, lubricants, antioxidants, and ultraviolet-light stabilizers. Why? What were the designers at the toy company thinking?

So much for trying to maintain a healthy environment, or even a healthy home. So much for peace, comfort, and safety. Something seems to be terribly wrong with this picture.

Now look at and feel the book in your hands.

This book is not a tree.

It is printed on a synthetic "paper" and bound into a book format developed by innovative book packager Charles Melcher of Melcher Media. Unlike the paper with which we are familiar, it does not use any wood pulp or cotton fiber but is made from plastic resins and inorganic fillers. This material is not only waterproof, extremely durable, and (in many localities) recyclable by conventional means; it is also a prototype for the book as a "technical nutrient," that is, as a product that can be broken down and circulated infinitely in industrial cycles—made and remade as "paper" or other products.

The tree, among the finest of nature's creations, plays a crucial and multifaceted role in our interdependent ecosystem. As such, it has been an important model and metaphor for our thinking, as you will discover. But also as such, it is not a fitting resource to use in producing so humble and transient a substance as paper. The use of an alternative material expresses our intention to evolve away from the use of wood fibers for paper as we seek more effective solutions. It represents one step toward a radically different approach to designing and producing the objects we use and enjoy, an emerging movement we see as the next industrial revolution. This revolution is founded on nature's surprisingly effective design principles, on human creativity and prosperity, and on respect, fair play, and goodwill. It has the power to transform both industry and environmentalism as we know them.

Toward a New Industrial Revolution

We are accustomed to thinking of industry and the environment as being at odds with each other, because conventional methods of extraction, manufacture, and disposal are destructive to the natural world. Environmentalists often characterize business as bad and industry itself (and the growth it demands) as inevitably destructive.

On the other hand, industrialists often view environmentalism as an obstacle to production and growth. For the environment to be healthy, the conventional attitude goes, industries must be regulated and restrained. For industries to fatten, nature cannot take precedence. It appears that these two systems cannot thrive in the same world.

The environmental message that "consumers" take from all this can be strident and depressing: Stop being so bad, so materialistic, so greedy. Do whatever you can, no matter how inconvenient, to limit your "consumption." Buy less, spend less, drive less, have fewer children—or none. Aren't the major environmental problems today—global warming, deforestation, pollution, waste—products of your decadent Western way of life? If you are going to help save the planet, you will have to make some sacrifices, share some resources, perhaps even go without. And fairly soon you must face a world of limits. There is only so much the Earth can take.

Sound like fun?

We have worked with both nature and commerce, and we don't think so.

One of us (Bill) is an architect, the other (Michael) is a chemist. When we met, you might say we came from opposite ends of the environmental spectrum.

Bill recalls:

I was influenced strongly by experiences I'd had abroad—first in Japan, where I spent my early childhood. I recall a sense of land and resources being scarce but also the beauty of traditional Japanese homes, with their paper walls and dripping gardens, their warm futons and steaming baths. I also remember quilted winter garments and farmhouses with thick walls of clay and straw that kept the interior warm in winter and cool in summer. Later, in college, I accompanied a professor of urban design to Jordan to develop housing for the Bedouin who were settling in the Jordan River valley. There I encountered an even greater scarcity of local resources—food, soil, energy, and especially water—but I was again struck by how simple and elegant good design could be, and how suited to locale. The tents of woven goat hair the Bedouin had used as nomads drew hot air up and out, creating not only shade but a breeze in their interiors. When it rained, the fibers swelled, and the structure became tight as a drum. It was portable and easily repaired: the fabric factory—the goats—followed the Bedouin around. This ingenious design, locally relevant, culturally rich, and using simple materials, contrasted sharply with the typical modern designs I had seen in my own country, designs that rarely made such good use of local material and energy flows.

When I returned to the States and entered graduate school, energy efficiency was the only real "environmental" topic considered by designers and architects. Interest in solar power had been piqued in the seventies when gas prices soared. I designed and built the first solar-heated house in Ireland (a measure of my ambition, there being very little sun in Ireland), which gave me a taste of the difficulties of applying universal solutions to local circumstances. Among the strategies experts suggested to me was building a huge rock storage bin to retain heat, which I discovered—after hauling thirty tons of rock—was redundant in an Irish house, with its thick masonry walls.

After graduate school, I apprenticed with a New York firm well known for its sensitive, socially responsible urban housing and then founded my own firm in 1981. In 1984 we were commissioned to design the offices of the Environmental Defense Fund, the first of the so-called green offices. I worked on indoor air quality, a subject almost no one had studied in depth. Of particular concern to us were volatile organic compounds, carcinogenic materials, and anything else in the paints, wall coverings, carpetings, floorings, and fixtures that might cause indoor air quality problems or multiple chemical sensitivity. With little or no research available, we turned to the manufacturers, who often told us the information was proprietary and gave us nothing beyond the vague safeguards in the material safety data sheets mandated by law. We did the best we could at the time. We used water-based paints. We tacked down carpet instead of gluing it. We provided thirty cubic feet per minute of fresh air per person instead of five. We had granite checked for radon. We used wood that was sustainably harvested. We tried to be less bad.

Most leading designers eschewed environmental concerns. Many environmentally minded designers applied environmental "solutions" in isolation, tacking new technology onto the same old model or coming up with giant solar collectors for people to live in that overheated in the summer. The resulting buildings were often ugly and obtrusive, and they were often not very effective. Even as architects and industrial designers began to embrace recycled or sustainable materials, they still dealt primarily with surfaces—with what looked good, what was easy to get, what they could afford.

I hoped for more. Two projects in particular inspired me to think seriously about my design intentions. In 1987 members of the Jewish community in New York asked me to design a proposal for a Holocaust memorial, a place where people could reflect. I visited Auschwitz and Birkenau to see what the worst of human intentions could accomplish: giant machines designed to eliminate human life. I realized that design is a signal of intention. What is the very best that designers can intend, I wondered, and how might a building manifest that intention? The second project was a proposal for a day-care center in Frankfurt, Germany, which again brought the issue of indoor air quality to the fore. What did it mean to design something that would be completely safe for children, particularly when safe building materials did not seem to exist?

I was tired of working hard to be less bad. I wanted to be involved in making buildings, even products, with completely positive intentions.

Michael's story:

I come from a family of literature and philosophy scholars, and turned to chemistry only out of sympathy for my high school chemistry teacher. (In the early 1970s Germany was engaged in political debate about the use of pesticides and other problematic chemicals, so I was able to justify it to my family as a meaningful pursuit.) I studied at universities where I could learn about environmental chemistry and was especially influenced by Professor Friedhelm Korte, who was instrumental in inventing "ecological chemistry." In 1978 I became one of the founding members of the Green Action Future Party. This became Germany's Green Party, and its primary goal was taking care of the environment.

Through my work with the Green Party, I created a name for myself among environmentalists. Greenpeace, which at the time was a group of activists with little formal background in science or environmental studies, asked me to work with them. I directed Greenpeace's chemistry department and helped the organization to protest more knowledgeably, but I soon realized that protest wasn't enough. We needed to develop a process for change. My turning point came after an action protesting a series of chemical spills by the big companies Sandoz and Ciba-Geigy: After a fire at Sandoz's huge factory was doused with firefighting chemicals that then ran into the Rhine, causing massive loss of river wildlife for more than a hundred miles, I coordinated a protest in which my colleagues and I chained ourselves to Ciba-Geigy smokestacks in Basel. When the activists came down two days later, Anton Schaerli, the director of the company, presented us with flowers and hot soup. Although he disagreed with our way of showing displeasure, he had been worried about us and wanted to hear what we had to say.

I explained that with Greenpeace's financing, I was about to start up an environmental chemistry research agency. I told him I planned to call it the Environmental Protection Enforcement Agency. The director was enthusiastic, and he suggested a slight modification in the name, from "Enforcement" to "Encouragement." It would be less hostile and more attractive to potential business clients, he said. I took his advice.

And so I became director of the EPEA, opening offices in several countries and continuing to develop a relationship with this large corporation. Partly in response to a request from Alex Krauer, chairman of Ciba-Geigy, I began to discover the rich experience of other cultures in working within nutrient flows, such as that of the Yanomamo of Brazil, who cremate their dead and put the ashes into a banana soup that the tribe eats at a celebratory feast. Many peoples believe in karma and reincarnation, an "upcycling" of the soul, if you will. These perspectives broadened my response to the problem of waste in the Western European tradition.

But it remained difficult for me to find other chemists who were interested in these matters at all, let alone had any experience in them. The formal study of chemistry still mostly excludes environmental issues, and science as a whole is more invested in research than in implementing strategies of change. The scientific community is usually paid to study problems, not solutions; indeed, finding a solution to the problem under study usually brings an end to funding for research. This puts an odd pressure on scientists, who, like everyone else, must make a living. Moreover, we scientists are trained in analysis rather than synthesis. I could tell you all about the components and potential negative effects of plasticizers, PVC, heavy metals, and many other harmful substances, which I learned about in my primary research. But my colleagues and I lacked a vision for putting this environmental knowledge to work within beautiful designs. My worldview was not one of abundance, creativity, prosperity, and change.

When I first met Bill, the environmentalists I knew were looking ahead to the upcoming 1992 Earth Summit, where the main agendas were sustainable development and global warming. Industry representatives would be there, and so would environmentalists. At the time, I still believed the two were destined to conflict. I was caught up in the notion that industry was bad, and environmentalism was ethically superior to it. I was concentrating on analyzing the often dangerous or questionable materials that went into everyday products like televisions, in hopes of devising a strategy to allow us to avoid the worst consequences of industrialism.

We met in 1991, when the EPEA held a reception at a rooftop garden in New York City to celebrate the opening of its first American offices. (The invitations were printed on biodegradable diapers, to highlight the fact that conventional disposable diapers were one of the largest single sources of solid waste in landfills.) We began talking about toxicity and design. Michael explained his idea of creating a biodegradable soda bottle with a seed implanted in it, which could be thrown on the ground after use to safely decompose and allow the seed to take root in the soil. There was music and dancing, and our discussion turned to another object of modern manufacture: the shoe. Michael joked that his guests were wearing "hazardous waste" on their feet, waste that was abrading as they spun on the rough surface of the roof, creating dust that people could inhale. He told how he had visited the largest chromium extraction factory in Europe—chromium is a heavy metal used in large-scale leather tanning processes—and noticed that only older men were working there, all of them in gas masks. The supervisor had explained that it took on average about twenty years for workers to develop cancer from chromium exposure, so the company had made the decision to allow only workers older than fifty to work with this dangerous substance.

There were other negative consequences associated with the conventional design of shoes, Michael pointed out. "Leather" shoes are actually a mixture of biological materials (the leather, which is biodegradable) and technical materials (the chromium and other substances, which have value for industries). According to current methods of manufacture and disposal, neither could be successfully retrieved after the shoe was discarded. From a material and ecological standpoint, the design of the average shoe could be much more intelligent. We discussed the idea of a sole coated with biodegradable materials, which could be detached after use. The rest of the shoe could be made of plastics and polymers that were not harmful, and which could be truly recycled into new shoes.

Incinerator smoke drifted from nearby rooftops as we discussed the fact that typical garbage, with its mixture of industrial and biological materials, was not designed for safe burning. Instead of banning burning, we wondered, why not manufacture certain products and packaging that could be safely burned after a customer is finished with them? We imagined a world of industry that made children the standard for safety. What about designs that, as Bill put it, "loved all the children, of all species, for all time"?

Traffic was increasing on the streets below, a true New York traffic jam, with blaring horns, angry drivers, and increasing disruption. In the early evening light, we imagined a silent car that could run without burning fossil fuels or emitting noxious fumes, and a city like a forest, cool and quiet. Everywhere we turned, we could see products, packaging, buildings, transportation, even whole cities that were poorly designed. And we could see that the conventional environmental approaches—

even the most well-intended and progressive ones—just didn't get it.

That initial meeting sparked an immediate interest in working together, and in 1991 we coauthored *The Hannover Principles*, design guidelines for the 2000 World's Fair that were issued at the World Urban Forum of the Earth Summit in 1992. Foremost among them was "Eliminate the concept of waste"—not reduce, minimize, or avoid waste, as environmentalists were then propounding, but eliminate the very concept, by design. We met in Brazil to see an early version of this principle in practice: a waste-processing garden that was in essence a giant intestine for its community, turning waste into food.

Three years later, we founded McDonough Braungart Design Chemistry. Bill maintained his architectural practice and Michael continued to head the EPEA in Europe, and both of us started teaching at universities. But now we had a focused way to begin to put our ideas into practice, to turn our work in chemical research, architecture, urban design, and industrial product and process design to the project of transforming industry itself. Since then, our design firms have worked with a wide range of corporate and institutional clients, including the Ford Motor Company, Herman Miller, Nike, and SC Johnson, and with a number of municipalities and research and educational institutions to implement the design principles we have evolved.

We see a world of abundance, not limits. In the midst of a great deal of talk about reducing the human ecological footprint, we offer a different vision. What if humans designed products and systems that celebrate an abundance of human creativity, culture, and productivity? That are so intelligent and safe, our species leaves an ecological footprint to delight in, not lament?

Consider this: all the ants on the planet, taken together, have a biomass greater than that of humans. Ants have been incredibly industrious for millions of years. Yet their productiveness nourishes plants, animals, and soil. Human industry has been in full swing for little over a century, yet it has brought about a decline in almost every ecosystem on the planet. Nature doesn't have a design problem. People do.

Chapter One

A Question of Design

In the spring of 1912, one of the largest moving objects ever created by human beings left Southampton, England, and began gliding toward New York. It appeared to be the epitome of its industrial age—a potent representation of technology, prosperity, luxury, and progress. It weighed 66,000 tons. Its steel hull stretched the length of four city blocks. Each of its steam engines was the size of a town house. And it was headed for a disastrous encounter with the natural world.

This vessel, of course, was the *Titanic*, a brute of a ship, seemingly impervious to the forces of the natural world. In the minds of the captain, the crew, and many of the passengers, nothing could sink it.

One might say that the *Titanic* was not only a product of the Industrial Revolution but remains an apt metaphor for the industrial infrastructure that revolution created. Like that famous ship, this infrastructure is powered by brutish and artificial sources of energy that are environmentally depleting. It pours waste into the water and smoke into the sky. It attempts to work by its own rules, which are contrary to those of nature. And although it may seem invincible, the fundamental flaws in its design presage tragedy and disaster.

A Brief History of the Industrial Revolution

Imagine that you have been given the assignment of designing the Industrial Revolution—retrospectively. With respect to its negative consequences, the assignment would have to read something like this:

Design a system of production that

- puts billions of pounds of toxic material into the air, water, and soil every year
- produces some materials so dangerous they will require constant vigilance by future generations
- results in gigantic amounts of waste
- puts valuable materials in holes all over the planet, where they can never be retrieved
- requires thousands of complex regulations—not to keep people and natural systems safe, but rather to keep them from being poisoned too quickly
- measures productivity by how few people are working
- creates prosperity by digging up or cutting down natural resources and then burying or burning them
- erodes the diversity of species and cultural practices.

Of course, the industrialists, engineers, inventors, and other minds behind the Industrial Revolution never intended such consequences. In fact, the Industrial Revolution as a whole was not really designed. It took shape gradually, as industrialists, engineers, and designers tried to solve problems and to take immediate advantage of what they considered to be opportunities in an unprecedented period of massive and rapid change.

It began with textiles in England, where agriculture had been the main occupation for centuries. Peasants farmed, the manor and town guilds provided food and goods, and industry consisted of craftspeople working individually as a side venture to farming. Within a few decades, this cottage industry, dependent on the craft of individual laborers for the production of small quantities of woolen cloth, was transformed into a mechanized factory system that churned out fabric—much of it now cotton instead of wool—by the mile.

This change was spurred by a quick succession of new technologies. In the mid-1700s cottage workers spun thread on spinning wheels in their homes, working the pedals with their hands and feet to make one thread at a time. The spinning jenny, patented in 1770, increased the number of threads from one to eight, then sixteen, then more. Later models would spin as many as eighty threads simultaneously. Other mechanized equipment, such as the water frame and the spinning mule, increased production levels at such a pace, it must have seemed something like Moore's Law (named for Gordon Moore, a founder of Intel), in which the processing speed of computer chips roughly doubles every eighteen months.

In preindustrial times, exported fabrics would travel by canal or sailing ships, which were slow and unreliable in poor weather, weighted with high duties and strict laws, and vulnerable to piracy. In fact, it was a wonder the cargo got to its destination at all. The railroad and the steamship allowed products to be moved more quickly and farther. By 1840 factories that had once made a thousand articles a week had the means and motivation to produce a thousand articles a day. Fabric workers grew too busy to farm and moved into towns to be closer to factories, where they and their families might work twelve or more hours a day. Urban areas spread, goods proliferated, and city populations increased. More, more, more—jobs, people, products, factories, businesses, markets—seemed to be the rule of the day.

Like all paradigm shifts, this one encountered resistance. Cottage workers afraid of losing work and Luddites (followers of Ned Ludd)—experienced cloth makers angry about the new machines and the unapprenticed workers who operated them—smashed labor-saving equipment and made life difficult for inventors, some of whom died outcast and penniless before they could profit from their new machines. Resistance touched not simply on technology but on spiritual and imaginative life. The Romantic poets articulated the growing difference between the rural, natural landscape and that of the city—often in despairing terms: "Citys . . . are nothing less than over grown prisons that shut out the world and all its beauties," wrote the poet John Clare. Artists and aesthetes like John Ruskin and William Morris feared for a civilization whose aesthetic sensibility and physical structures were being reshaped by materialistic designs.

There were other, more lasting problems. Victorian London was notorious for having been "the great and dirty city," as Charles Dickens called it, and its unhealthy environment and suffering underclasses became hallmarks of the burgeoning industrial city. London air was so grimy from airborne pollutants, especially emissions from burning coal, that people would change their cuffs and collars at the end of the day (behavior that would be repeated in Chattanooga during the 1960s, and even today in Beijing or Manila). In early factories and other industrial operations, such as mining, materials were considered expensive, but people were often considered cheap. Children as well as adults worked for long hours in deplorable conditions.

But the general spirit of early industrialists—and of many others at the time—was one of great optimism and faith in the progress of humankind. As industrialization boomed, other institutions emerged that assisted its rise: commercial banks, stock exchanges, and the commercial press all opened further employment opportunities for the new middle class and tightened the social network around economic growth. Cheaper products, public transportation, water distribution and sanitation, waste collection, laundries, safe housing, and other conveniences gave people, both rich and poor, what appeared to be a more equitable standard of living. No longer did the leisure classes alone have access to all the comforts.

The Industrial Revolution was not planned, but it was not without a motive. At bottom it was an economic revolution, driven by the desire for the acquisition of capital. Industrialists wanted to make products as efficiently as possible and to get the greatest volume of goods to the largest number of people. In most industries, this meant shifting from a system of manual labor to one of efficient mechanization.

Consider cars. In the early 1890s the automobile (of European origin) was made to meet a customer's specifications by craftspeople who were usually independent contractors. For example, a machine-tool company in Paris, which happened to be the leading manufacturer of cars at the time, produced only several hundred a year. They were luxury items, built slowly and carefully by hand. There was no standard system of measuring and gauging parts, and no way to cut hard steel, so parts were created by different contractors, hardened under heat (which often altered dimensions), and individually filed down to fit the hundreds of other parts in the car. No two were alike, nor could they be.

Henry Ford worked as an engineer, a machinist, and a builder of race cars (which he himself raced) before founding the Ford Motor Company in 1903. After producing a number of early vehicles, Ford realized that to make cars for the modern American worker—not just for the wealthy—he would need to manufacture vehicles cheaply and in great quantities. In 1908 his company began producing the legendary Model T, the "car for the great multitude" that Ford had dreamed of, "constructed of the best materials, by the best men to be hired, after the simplest designs that modern engineering can devise . . . so low in price that no man making a good salary will be unable to own one."

In the following years, several aspects of manufacturing meshed to achieve this goal, revolutionizing car production and rapidly increasing levels of efficiency. First, centralization: in 1909 Ford announced that the company would produce only Model T's and in 1910 moved to a much larger factory that would use electricity for its power and gather a number of production processes under one roof. The most famous of Ford's innovations is the moving assembly line. In early production, the engines, frames, and bodies of the cars were assembled separately, then brought together for final assembly by a group of workmen. Ford's innovation was to bring "the materials to the man," instead of "the man to the materials." He and his engineers developed a moving assembly line based on the ones used in the Chicago beef industry: it carried materials to workers and, at its most efficient, enabled each of them to repeat a single operation as the vehicle moved down the line, reducing overall labor time dramatically.

This and other advances made possible the mass production of the universal car, the Model T, from a centralized location, where many vehicles were assembled at once. Increasing efficiency pushed costs of the Model T down (from $850 in 1908 to $290 in 1925), and sales skyrocketed. By 1911, before the introduction of the assembly line, sales of the Model T had totaled 39,640. By 1927, total sales reached fifteen million.

The advantages of standardized, centralized production were manifold. Obviously, it could bring greater, quicker affluence to industrialists. On another front, manufacturing was viewed as what Winston Churchill referred to as "the arsenal of democracy," because the productive capacity was so huge, it could (as in the two world wars) produce an undeniably potent response to war conditions. Mass production had another democratizing aspect: as the Model T demonstrated, when prices of a previously unattainable item or service plummeted, more people had access to it. New work opportunities in factories improved standards of living, as did wage increases. Ford himself assisted in this shift. In 1914, when the prevailing salary for factory workers was $2.34 a day, he hiked it to $5, pointing out that cars cannot buy cars. (He also reduced the hours of the workday from nine to eight.) In one fell swoop, he actually created his own market, and raised the bar for the entire world of industry.

Viewed from a design perspective, the Model T epitomized the general goal of the first industrialists: to make a product that was desirable, affordable, and operable by anyone, just about anywhere; that lasted a certain amount of time (until it was time to buy a new one); and that could be produced cheaply and quickly. Along these lines, technical developments centered on increasing "power, accuracy, economy, system, continuity, speed," to use the Ford manufacturing checklist for mass production.

For obvious reasons, the design goals of early industrialists were quite specific, limited to the practical, profitable, efficient, and linear. Many industrialists, designers, and engineers did not see their designs as part of a larger system, outside of an economic one. But they did share some general assumptions about the world.

"Those Essences Unchanged by Man"

Early industries relied on a seemingly endless supply of natural "capital." Ore, timber, water, grain, cattle, coal, land—these were the raw materials for the production systems that made goods for the masses, and they still are today. Ford's River

Rouge plant epitomized the flow of production on a massive scale: huge quantities of iron, coal, sand, and other raw materials entered one side of the facility and, once inside, were transformed into new cars. Industries fattened as they transformed resources into products. The prairies were overtaken for agriculture, and the great forests were cut down for wood and fuel. Factories situated themselves near natural resources for easy access (today a prominent window company is located in a place that was originally surrounded by giant pines, used for the window frames) and beside bodies of water, which they used both for manufacturing processes and to dispose of wastes.

In the nineteenth century, when these practices began, the subtle qualities of the environment were not a widespread concern. Resources seemed immeasurably vast. Nature itself was perceived as a "mother earth" who, perpetually regenerative, would absorb all things and continue to grow. Even Ralph Waldo Emerson, a prescient philosopher and poet with a careful eye for nature, reflected a common belief when, in the early 1830s, he described nature as "essences unchanged by man; space, the air, the river, the leaf." Many people believed there would always be an expanse that remained unspoiled and innocent. The popular fiction of Rudyard Kipling and others evoked wild parts of the world that still existed and, it seemed, always would.

At the same time, the Western view saw nature as a dangerous, brutish force to be civilized and subdued. Humans perceived natural forces as hostile, so they attacked back to exert control. In the United States, taming the frontier took on the power of a defining myth, and "conquering" wild, natural places was recognized as a cultural—even spiritual—imperative.

Today our understanding of nature has dramatically changed. New studies indicate that the oceans, the air, the mountains, and the plants and animals that inhabit them are more vulnerable than early innovators ever imagined. But modern industries still operate according to paradigms that developed when humans had a very different sense of the world. Neither the health of natural systems, nor an awareness of their delicacy, complexity, and interconnectedness, have been part of the industrial design agenda. At its deepest foundation, the industrial infrastructure we have today is linear: it is focused on making a product and getting it to a customer quickly and cheaply without considering much else.

To be sure, the Industrial Revolution brought a number of positive social changes. With higher standards of living, life expectancy greatly increased. Medical care and education greatly improved and became more widely available. Electricity, telecommunications, and other advances raised comfort and convenience to a new level. Technological advances brought the so-called developing nations enormous benefits, including increased productivity of agricultural land and vastly increased harvests and food storage for growing populations.

But there were fundamental flaws in the Industrial Revolution's design. They resulted in some crucial omissions, and devastating consequences have been handed down to us, along with the dominant assumptions of the era in which the transformation took shape.

From Cradle to Grave

Imagine what you would come upon today at a typical landfill: old furniture, upholstery, carpets, televisions, clothing, shoes, telephones, computers, complex products, and plastic packaging, as well as organic materials like diapers, paper, wood, and food wastes. Most of these products were made from valuable materials that required effort and expense to extract and make, billions of dollars' worth of material assets. The biodegradable materials such as food matter and paper actually have value too—they could decompose and return biological nutrients to the soil. Unfortunately, all of these things are heaped in a landfill, where their value is wasted. They are the ultimate products of an industrial system that is designed on a linear, one-way *cradle-to-grave* model. Resources are extracted, shaped into products, sold, and eventually disposed of in a "grave" of some kind, usually a landfill or incinerator. You are probably familiar with the end of this process because you, the customer, are responsible for dealing with its detritus. Think about it: you may be referred to as a consumer, but there is very little that you actually consume—some food, some liquids. Everything else is designed for you to throw away when you are finished with it. But where is "away"? Of course, "away" does not really exist. "Away" has gone away.

Cradle-to-grave designs dominate modern manufacturing. According to some accounts more than 90 percent of materials extracted to make durable goods in the United States become waste almost immediately. Sometimes the product itself scarcely lasts longer. It is often cheaper to buy a new version of even the most expensive appliance than to track down someone to repair the original item. In fact, many products are designed with "built-in obsolescence," to last only for a certain period of time, to allow—to encourage—the customer to get rid of the thing and buy a new model. Also, what most people see in their garbage cans is just the tip of a material iceberg; the product itself contains on average only 5 percent of the raw materials involved in the process of making and delivering it.

One Size Fits All

Because the cradle-to-grave model underlying the design assumptions of the Industrial Revolution was not called into question, even movements that were formed ostensibly in opposition to that era manifested its flaws. One example has been the push to achieve universal design solutions, which emerged as a leading design strategy in the last century. In the field of architecture, this strategy took the form of the International Style movement, advanced during the early decades of the twentieth century by figures such as Ludwig Mies van der Rohe, Walter Gropius, and Le Corbusier, who were reacting against Victorian-era styles. (Gothic cathedrals were still being proposed and built.) Their goals were social as well as aesthetic. They wanted to globally replace unsanitary and inequitable housing—fancy, ornate places for the rich; ugly, unhealthy places for the poor—with clean, minimalist, affordable buildings unencumbered by distinctions of wealth or class. Large sheets of glass, steel, and concrete, and cheap transportation powered by fossil fuels, gave engineers and architects the tools for realizing this style anywhere in the world.

Today the International Style has evolved into something less ambitious: a bland, uniform structure isolated from the particulars of place—from local culture, nature, energy, and material flows. Such buildings reflect little if any of a region's distinctness or style. They often stand out like sore thumbs from the surrounding landscape, if they leave any of it intact around their "office parks" of asphalt and concrete. The interiors are equally uninspiring. With their sealed windows, constantly humming air conditioners, heating systems, lack of daylight and fresh air, and uniform fluorescent lighting, they might as well have been designed to house machines, not humans.

The originators of the International Style intended to convey hope in the "brotherhood" of humankind. Those who use the style today do so because it is easy and cheap and makes architecture uniform in many settings. Buildings can look and work the same anywhere, in Reykjavík or Rangoon.

In product design, a classic example of the universal design solution is mass-produced detergent. Major soap manufacturers design one detergent for all parts of the United States or Europe, even though water qualities and community needs differ. For example, customers in places with soft water, like the Northwest, need only small amounts of detergent. Those where the water is hard, like the Southwest, need more. But detergents are designed so they will lather up, remove dirt, and kill germs efficiently the same way anywhere in the world—in hard, soft, urban, or spring water, in water that flows into fish-filled streams and water channeled to sewage treatment plants. Manufacturers just add more chemical force to wipe out the conditions of circumstance. Imagine the strength a detergent must have to strip day-old grease from a greasy pan. Now imagine what happens when that detergent comes into contact with the slippery skin of a fish or the waxy coating of a plant. Treated and untreated effluents as well as runoff are released into lakes, rivers, and oceans. Combinations of chemicals, from household detergents, cleansers, and medicines along with industrial wastes, end up in sewage effluents, where they have been shown to harm aquatic life, in some cases causing mutations and infertility.

To achieve their universal design solutions, manufacturers design for a *worst-case scenario*; they design a product for the worst possible circumstance, so that it will always operate with the same efficacy. This aim guarantees the largest possible market for a product. It also reveals human industry's peculiar relationship to the natural world, since designing for the worst case at all times reflects the assumption that nature is the enemy.

Brute Force

If the first Industrial Revolution had a motto, we like to joke, it would be "If brute force doesn't work, you're not using enough of it." The attempt to impose universal design solutions on an infinite number of local conditions and customs is one manifestation of this principle and its underlying assumption, that nature should be overwhelmed; so is the application of the chemical brute force and fossil fuel energy necessary to make such solutions "fit."

All of nature's industry relies on energy from the sun, which can be viewed as a form of current, constantly renewing income. Humans, by contrast, extract and burn fossil fuels such as coal and petrochemicals that have been deposited deep below the Earth's surface, supplementing them with energy produced through waste-incineration processes and nuclear reactors that create additional problems. They do this with little or no attention to harnessing and maximizing local natural energy flows. The standard operating instruction seems to be "If too hot or too cold, just add more fossil fuels."

You are probably familiar with the threat of global warming brought about by the buildup of heat-trapping gases (such as carbon dioxide) in the atmosphere due to human activities. Increasing global temperatures result in global climate change and shifts of existing climates. Most models predict more severe weather: hotter hots, cooler colds, and more intense storms, as global thermal contrasts grow more extreme. A warmer atmosphere draws more water from oceans, resulting in bigger, wetter, more frequent storms, rises in sea level, shifts in seasons, and a chain of other climatic events.

The reality of global warming has gained currency not only among environmentalists but among industry leaders. But global warning is not the sole reason to rethink our reliance on the "brute force" approach to energy. Incinerating fossil fuels contributes particulates—microscopic particles of soot—to the environment, where they are known to cause respiratory and other health problems. Regulations for airborne pollutants known to threaten health are growing more severe. As new regulations, based on mounting research about the health threats of airborne toxins resulting from incinerating fossil fuels, are implemented, industries invested solely in continuing the current system will be at a serious disadvantage.

Even beyond these important issues, brute force energy doesn't make good sense as a dominant strategy over the long term. You wouldn't want to depend on savings for all of your daily expenditures, so why rely on savings to meet all of humanity's energy needs? Clearly, over the years petrochemicals will become harder (and more expensive) to get, and drilling in pristine places for a few million more drums of oil isn't going to solve that problem. In a sense, finite sources of energy, such as petrochemicals derived from fossil fuels, can be seen as a nest egg, something to be preserved for emergencies, then used sparingly—in certain medical situations, for example. For the majority of our simple energy needs, humans could be accruing a great deal of current solar income, of which there is plenty: thousands of times the amount of energy needed to fuel human activities hits the surface of the planet every day in the form of sunlight.

A Culture of Monoculture

Under the existing paradigm of manufacturing and development, diversity—an integral element of the natural world—is typically treated as a hostile force and a threat to design goals.

Brute force and universal design approaches to typical development tend to overwhelm (and ignore) natural and cultural diversity, resulting in less variety and greater homogeneity.

Consider the process of building a typical universal house. First builders scrape away everything on the site until they reach a bed of clay or undisturbed soil. Several machines then come in and shape the clay to a level surface. Trees are felled, natural flora and fauna are destroyed or frightened away, and the generic mini McMansion or modular home rises with little regard for the natural environment around it—ways the sun might come in to heat the house during the winter, which trees might protect it from wind, heat, and cold, and how soil and water health can be preserved now and in the future. A two-inch carpet of a foreign species of grass is placed over the rest of the lot.

The average lawn is an interesting beast: people plant it, then douse it with artificial fertilizers and dangerous pesticides to make it grow and to keep it uniform—all so that they can hack and mow what they encouraged to grow. And woe to the small yellow flower that rears its head!

Rather than being designed around a natural and cultural landscape, most modern urban areas simply grow, as has often been said, like a cancer, spreading more and more of themselves, eradicating the living environment in the process, blanketing the natural landscape with layers of asphalt and concrete.

Conventional agriculture tends to work along these same lines. The goal of a midwestern commercial corn operation is to produce as much corn as possible with the least amount of trouble, time, and expense—the Industrial Revolution's first design goal of maximum efficiency. Most conventional operations today focus on highly specialized, hybridized, and perhaps genetically modified species of corn. They develop a monocultural landscape that appears to support only one particular crop that's likely not even a true species but some over-hybridized cultivar. Planters remove other species of plant life using tillage, which leads to massive soil erosion from wind and water, or no-till farming, which requires massive applications of herbicide. Ancient strains of corn are lost because their output does not meet the demands of modern commerce.

On the surface, these strategies seem reasonable to modern industry and even to "consumers," but they harbor both underlying and overlying problems. Elements that are removed from the ecosystem to make the operation yield more grain more quickly (that is, to make it more efficient) would otherwise actually provide benefits to farming. The plants removed by tillage, for example, could have helped to prevent erosion and flooding and to stabilize and rebuild soil. They would have provided habitat for insects and birds, some of them natural enemies of crop pests. Now, as pests grow resistant to pesticide, their numbers increase because their natural enemies have been wiped out.

Pesticides, as typically designed, are a perennial cost both to farmers and to the environment and represent a less than mindful use of chemical brute force. Although chemical companies warn farmers to be careful with pesticides, they benefit when more of them are sold. In other words, the companies are unintentionally invested in profligacy with—even the mishandling of—their products, which can result in contamination of the soil, water, and air.

In such an artificially maintained system, where the natural enemies of pests and some of the nutrient-cycling plants and organisms have been eliminated, more chemical brute force (pesticides, fertilizers) must be applied to keep the system commercially stable. Soil is depleted of nutrients and saturated with chemicals. People may not want to live too close to the operation because they fear chemical runoff. Rather than being an aesthetic and cultural delight, modern agriculture becomes a terror and a fright to local residents who want to live and raise families in a healthy setting. While the economic payoff immediately rises, *the overall quality of every aspect of this system is actually in decline.*

The problem here is not agriculture per se but the narrowly focused goals of the operation. The single-minded cultivation of one species drastically reduces the rich network of "services" and side effects in which the entire ecosystem originally engaged. To this day, conventional agriculture is still, as scientists Paul and Anne Ehrlich and John Holdren said several decades ago, "a simplifier of ecosystems, replacing relatively complex natural biological communities with relatively simple man-made ones based on a few strains of crops." These simple systems cannot survive on their own. Ironically, simplification necessitates even more brute force for the system to achieve its design goals. Take away the chemicals and the modern modes of agricultural control, and the crops would languish (until, that is, diverse species gradually crept back, returning complexity to the ecosystem).

Activity Equals Prosperity

An interesting fact: the 1991 Exxon *Valdez* oil spill actually increased Alaska's gross domestic product. The Prince William Sound area was registered as economically more prosperous because so many people were trying to clean up the spill. Restaurants, hotels, shops, gas stations, and stores all experienced an upward blip in economic exchange.

The GDP takes only one measure of progress into account: activity. Economic activity. But what sensible person would call the effects of an oil spill progress? By some accounts, the *Valdez* accident led to the death of more wildlife than any other human-engineered environmental disaster in U.S. history. According to a 1999 government report, only two of the twenty-three animal species affected by the spill recovered. Its impact on fish and wildlife continues today with tumors, genetic damage, and other effects. The spill led to losses of cultural wealth, including five state parks, four state critical-habitat areas, and a state game sanctuary. Important habitats for fish spawning and rearing were damaged, which may have led to the 1993 decimation of the Prince William Sound's Pacific herring population (perhaps because of a viral infection due to oil exposure). The spill took a significant toll on fishermen's income, not to mention the less measurable effects on morale and emotional health.

The GDP as a measure of progress emerged during an era when natural resources still seemed unlimited and "quality of life" meant high economic standards of living. But if prosperity is judged only by increased economic activity, then car accidents, hospital visits, illnesses (such as cancer), and toxic spills are all signs of prosperity. Loss of resources, cultural depletion, negative social and environmental effects, reduction of quality of life—these ills can all be taking place, an entire region can be in decline, yet they are negated by a simplistic economic figure that says economic life is good. Countries all over the world are trying to boost their level of economic activity so they, too, can grab a share of the "progress" that measurements like the GDP propound. But in the race for economic progress, social activity, ecological impact, cultural activity, and long-term effects can be overlooked.

Crude Products

The design intention behind the current industrial infrastructure is to make an attractive product that is affordable, meets regulations, performs well enough, and lasts long enough to meet market expectations. Such a product fulfills the manufacturer's desires and some of the customers' expectations as well. But from our perspective, products that are not designed particularly for human and ecological health are unintelligent and inelegant—what we call *crude products*.

For example, the average mass-produced piece of polyester clothing and a typical water bottle both contain antimony, a toxic heavy metal known to cause cancer under certain circumstances. Let's put aside for the moment the issues of whether this substance represents a specific danger to the user. The question we would pose as designers is: Why is it there? Is it necessary? Actually, it is not necessary: antimony is a current catalyst in the polymerization process and is not necessary for polyester production. What happens when this discarded product is "recycled" (that is, downcycled) and mixed with other materials? What about when it is burned along with other trash as cooking fuel, a common practice in developing countries? Incineration makes the antimony bioavailable—that is, available for breathing. If polyester might be used for fuel, we need polyesters that can be safely burned.

That polyester shirt and that water bottle are both examples of what we call *products plus*: as a buyer you got the item or service you wanted, *plus* additives that you didn't ask for and didn't know were included and that may be harmful to you and your loved ones. (Maybe shirt labels should read: *Product contains toxic dyes and catalysts. Don't work up a sweat or they will leach onto your skin.*) Moreover, these extra ingredients may not be necessary to the product itself.

Since 1987 we have been studying various products from major manufacturers, ordinary things such as a computer mouse, an electric shaver, a popular handheld video game, a hair dryer, and a portable CD player. We found that during use they all off-gassed teratogenic and/or carcinogenic compounds—substances known to have a role in causing birth defects and cancer. An electric hand mixer emitted chemical gases that got trapped in the oily butter molecules of the cake batter and ended up in the cake. So be careful—you might unintentionally be eating your appliances.

Why does this happen? The reason is that high-tech products are usually composed of low-quality materials—that is, cheap plastics and dyes—globally sourced from the lowest-cost provider, which may be halfway around the world. This means that even substances banned for use in the United States and Europe can reach this country via products and parts made elsewhere. So, for example, the carcinogen benzene, banned for use as a solvent in American factories, can be shipped here in rubber parts that were manufactured in developing countries that have not banned it. They can be assembled into, say, your treadmill, which will then emit the "banned" substance as you exercise.

The problem intensifies when parts from numerous countries are assembled into one product, as is often the case with high-tech items such as electronic equipment and appliances. Manufacturers do not necessarily keep track of—nor are they required to know—what exactly is in all of these parts. An exercise machine assembled in the United States may contain rubber belts from Malaysia, chemicals from Korea, motors from China, adhesives from Taiwan, and wood from Brazil.

How do these crude products affect you? They produce poor indoor air quality, for one thing. Combined in the workplace or home, crude products—whether appliances, carpets, wallpaper adhesives, paints, building materials, insulation, or anything else—make the average indoor air more contaminated than outdoor air. One study of household contaminants found that more than half of the households showed concentrations of seven toxic chemicals that are known to cause cancer in animals and are suspected to cause cancer in humans at levels higher than those that would "trigger a formal risk assessment for residential soil at a Superfund site." Allergies, asthma, and "sick building syndrome" are on the rise. Yet legislation establishing mandatory standards for indoor air quality is practically nonexistent.

Even products ostensibly designed for children can be crude products. An analysis of a child's swim wings, made from polyvinyl chloride (PVC), showed that they off-gassed potentially harmful substances—including, under heat, hydrochloric acid. Other harmful substances, like the plasticizing phthalates, may be ingested through contact. This scenario is particularly alarming in a swimming pool, since a child's skin, ten times thinner than the skin of an adult, gets wrinkled when wet—the ideal condition for absorbing toxins. Once again, in purchasing swim wings, you've inadvertently purchased a "product plus": you got the flotation device you wanted for your child *plus* unasked-for toxins—not a great bargain, and surely not what the manufacturers had in mind when they created this child-safety device.

You may be saying to yourself, "I certainly don't know any children who have gotten sick from a plastic float or pool!" But rather than a readily identifiable illness, some people develop an allergy, or multiple chemical sensitivity syndrome, or asthma, or they just do not feel well, without knowing exactly why. Even if we experience no immediate ill effects, coming into constant contact with carcinogens like benzene and vinyl chloride may be unwise.

Think of it this way. Everyone's body is subjected to stress, from both internal and external sources. These stresses may take the form of cancer cells that are naturally produced by the body—by some accounts, as many as twelve cells a day—exposure to heavy metals and other pathogens, and so on. The immune system is capable of handling a certain amount of stress. Simplistically speaking, you could picture those stressors as balls your immune system is juggling. Ordinarily, the juggler is skillful enough to keep those balls in the air. That is, the immune system catches and destroys those ten or twelve cells. But the more balls in the air—the more the body is besieged by all kinds of environmental toxins, for example—the greater the probability that it will drop the ball, that a replicating cell will make a mistake. It would be very hard to say which molecule or factor was the one that pushed a person's system over the edge. But why not remove negative stressors, especially since people don't want or need them?

Some industrial chemicals produce a second effect, more insidious than causing stress: they weaken the immune system. This is like tying one of the juggler's hands behind his back, which makes it much harder for him to catch the cancer cells before they cause problems. The deadliest chemicals both destroy the immune system *and* damage cells. Now you have a one-handed juggler struggling to keep an increasing number of balls in the air. Will he continue to perform with accuracy and grace? Why take the risk that he won't? Why not look for opportunities to strengthen the immune system, not challenge it?

We've focused on cancer here, but these compounds may have other effects that science has yet to discover. Consider endocrine disrupters, which were unheard of a decade ago but are now known to be among the most damaging chemical compounds for living organisms. Of the approximately eighty thousand defined chemical substances and technical mixes that are produced and used by industries today (each of which has five or more by-products), only about three thousand so far have been studied for their effects on living systems.

It may be tempting to try to turn back the clock. Yet the next industrial revolution will not be about returning to some idealized, preindustrial state in which, for example, all textiles are made from natural fibers. Certainly at one time fabrics were biodegradable and unwanted pieces could be tossed on the ground to decompose or even be safely burned as fuel. But the natural materials to meet the needs of our current population do not and cannot exist. If several billion people want natural-fiber blue jeans dyed with natural dyes, humanity will have to dedicate millions of acres to the cultivation of indigo and cotton plants just to satisfy the demand—acres that are needed to produce food. In addition, even "natural" products are not necessarily healthy for humans and the environment. Indigo contains mutagens and, as typically grown in monocultural practices, depletes genetic diversity. You want to change your jeans, not your genes. Substances created by nature can be extremely toxic; they were not specifically designed by evolution for our use. Even something as benign and necessary as clean drinking water can be lethal if you are submerged in it for more than a couple of minutes.

A Strategy of Tragedy, or a Strategy of Change?

Today's industrial infrastructure is designed to chase economic growth. It does so at the expense of other vital concerns, particularly human and ecological health, cultural and natural richness, and even enjoyment and delight. Except for a few generally known positive side effects, most industrial methods and materials are unintentionally depletive.

Yet just as industrialists, engineers, designers, and developers of the past did not intend to bring about such devastating effects, those who perpetuate these paradigms today surely do not intend to damage the world. The waste, pollution, crude products, and other negative effects that we have described are not the result of corporations doing something morally wrong. They are the consequence of outdated and unintelligent design.

Nevertheless, the damage is certain and severe. Modern industries are chipping away at some of the basic achievements that industrialization brought about. Food stocks, for example, have increased so that more children are fed, but more children go to bed hungry as well. But even if well-fed children are regularly exposed to substances that can lead to genetic mutations, cancer, asthma, allergies, and other complications from industrial contamination and waste, then what has been achieved? Poor design on such a scale reaches far beyond our own life span. It perpetrates what we call *intergenerational remote tyranny*—our tyranny over future generations through the effects of our actions today.

At some point a manufacturer or designer decides, "We can't keep doing this. We can't keep supporting and maintaining this system." At some point they will decide that they would prefer to leave behind a positive design legacy. But when is that point?

We say that point is today, and negligence starts tomorrow.

Once you understand the destruction taking place, unless you do something to change it, even if you never intended to cause such destruction, you become involved in a strategy of tragedy. You can continue to be engaged in that strategy of tragedy, or you can design and implement a *strategy of change*.

Perhaps you imagine that a viable strategy for change already exists. Aren't a number of "green," "environmental," and "eco-efficient" movements already afoot? The next chapter takes a closer look at these movements and the solutions they offer.

Chapter Two

Why Being "Less Bad" Is No Good

The drive to make industry less destructive goes back to the earliest stages of the Industrial Revolution, when factories were so destructive and polluting that they had to be controlled in order to prevent immediate sickness and death. Since then the typical response to industrial destruction has been to find a less bad approach. This approach has its own vocabulary, with which most of us are familiar: *reduce, avoid, minimize, sustain, limit, halt*. These terms have long been central to environmental agendas, and they have become central to most of the environmental agendas taken up by industry today.

One early dark messenger was Thomas Malthus, who warned at the end of the eighteenth century that humans would reproduce exponentially, with devastating consequences for humankind. Malthus's position was unpopular during the explosive excitement of early industry, when much was made of humanity's potential for good, when its increasing ability to mold the earth to its own purposes was seen as largely constructive; and when even population growth was viewed as a boon. Malthus envisioned not great, gleaming advancement but darkness, scarcity, poverty, and famine. His *Population: The First Essay*, published in 1798, was framed as a response to essayist and utopian William Godwin, who often espoused man's "perfectibility." "I have read some of the speculations on the perfectibility of man and of society with great pleasure," Malthus wrote. "I have been warmed and delighted with the enchanting picture which they hold forth." But, he concluded, "The power of population is so superior to the power in the earth to produce subsistence for man, that premature death must in some shape or other visit the human race." Because of his pessimism (and his suggestion that people should have less sex), Malthus became a cultural caricature. Even now his name is equated with a Scrooge-like attitude toward the world.

While Malthus was making his somber predictions about human population and resources, others were noticing changes in nature (and spirit) as industry spread. English Romantic writers such as William Wordsworth and William Blake described the spiritual and imaginative depth that nature could inspire, and they spoke out against an increasingly mechanistic urban society that was turning even more of its attention toward getting and spending. The Americans George Perkins Marsh, Henry David Thoreau, John Muir, Aldo Leopold, and others continued this literary tradition into the nineteenth and twentieth centuries and in the New World. From the Maine woods, Canada, Alaska, the Midwest, and the Southwest, these voices from the wilderness preserved in language the landscape they loved, lamented its destruction, and reaffirmed the belief that, as Thoreau famously put it, "in Wildness is the preservation of the world." Marsh was one of the first to understand man's capacity to wreak lasting destruction on the environment, and Leopold anticipated some of the feelings of guilt that characterize much environmentalism today:

> When I submit these thoughts to a printing press, I am helping cut down the woods. When I pour cream in my

> coffee, I am helping to drain a marsh for cows to graze, and to exterminate the birds of Brazil. When I go birding or hunting in my Ford, I am devastating an oil field, and re-electing an imperialist to get me rubber. Nay more: when I father more than two children I am creating an insatiable need for more printing presses, more cows, more coffee, more oil, to supply which more birds, more trees, and more flowers will either be killed, or . . . evicted from their several environments.

Some of these men helped to form conservation societies, such as the Sierra Club and the Wilderness Society, to preserve wilderness and keep it untouched by industrial growth. Their writings inspired new generations of environmentalists and nature lovers, and they still do.

But it wasn't until the publication of Rachel Carson's *Silent Spring* in 1962 that this romantic strain of wilderness appreciation merged with a scientific basis for concern. Up until that point, environmentalism meant protesting the obvious damage—deforestation, mining destruction, factory pollution, and other visible changes—and seeking to conserve especially appreciated landscapes, like the White Mountains of New Hampshire or Yosemite in California. Carson pointed out something more insidious; she imagined a landscape in which no birds sang, and moved on to explain that human-made chemicals—particularly pesticides such as DDT—were devastating the natural world.

Although it took almost a decade, *Silent Spring* led to the banning of DDT in the United States and Germany and sparked a continuing controversy about the dangers of industrial chemicals. It influenced scientists and politicians to take up the cause and to form groups such as Environmental Defense, the Natural Resources Defense Council, the World Wildlife Federation, and BUND (the German Federation for Environmental and Nature Conservation). Environmentalists were no longer interested simply in preservation but in monitoring and reducing toxins. Declining wilderness and diminishing resources merged with pollution and toxic waste as the major realms of concern.

Malthus's legacy continued to hold strong. Shortly after *Silent Spring*, in 1968, Paul Ehrlich, a pioneer of modern environmentalism and an eminent biologist working at Stanford, published an alarm of Malthusian proportions, *The Population Bomb*, in which he declared that the 1970s and 1980s would be a dark era of resource shortages and famine, during which "hundreds of millions of people will starve to death." He also pointed out humans' habit of "using the atmosphere as a garbage dump." "Do we want to keep it up and find out what will happen?" he asked. "What do we gain by playing 'environmental roulette'?"

In 1984 Ehrlich and his wife, Anne, followed up the first book with another, *The Population Explosion*. In this second warning to humanity, they asserted, "Then the fuse was burning; now the population bomb has detonated." Primary among "the underlying causes of our planet's unease," the two posited, "is the overgrowth of the human population and its impacts on both ecosystems and human communities." Their first chapter is entitled "Why Isn't Everyone as Scared as We Are?" and their parting suggestion for humanity begins with two urgent suggestions: "Halt human population growth as quickly and humanely as possible," and "Convert the economic system from one of growthism to one of sustainability, lowering per-capita consumption."

The association of growth with negative consequences has become a major theme of environmentalists in the modern age. In 1972, between the publication of the Ehrlichs' first and second warnings, Donella and Dennis Meadows and the Club of Rome (a group of international business, state, and scientific leaders) published another serious warning, *The Limits to Growth*. The authors noted that resources were plummeting due to population growth and destructive industry and concluded, "If the present growth trends in world population, industrialization, pollution, food production, and resource depletion continue unchanged, the limits to growth on this planet will be reached sometime within the next one hundred years. The most probable result will be a sudden and uncontrollable decline in both population and industrial capacity." Twenty years later a follow-up, *Beyond the Limits*, concluded with more warnings: "Minimize the use of nonrenewable resources." "Prevent the erosion of renewable resources." "Use all resources with maximum efficiency." "Slow and eventually stop exponential growth of population and physical capital."

In 1973 Fritz Schumacher's *Small Is Beautiful: Economics as If People Mattered* tackled the issue of growth from a philosophical vantage point. "The idea of unlimited economic growth," he wrote, "more and more until everybody is saturated with wealth, needs to be seriously questioned." In addition to advocating small-scale, nonviolent technologies that would "reverse the destructive trends now threatening us all," Schumacher posited that people must make a serious shift in what they consider to be wealth and progress: "Ever-bigger machines, entailing ever-bigger concentrations of economic power and exerting ever-greater violence against the environment, do not represent progress: they are a denial of wisdom." Real wisdom, he claimed, "can be found only inside oneself," enabling one to "see the hollowness and fundamental unsatisfactoriness of a life devoted primarily to the pursuit of material ends."

At the same time that these environmentalists were issuing important warnings, others were suggesting ways consumers could reduce their negative impact on the environment. A recent version of this message is found in Robert Lilienfeld and William Rathje's 1998 *Use Less Stuff: Environmental Solutions for Who We Really Are*. Consumers must take the lead in reducing negative environmental impact, the authors argue: "The simple truth is that all of our major environmental concerns are either caused by, or contribute to, the ever-increasing consumption of goods and services." This devouring impulse in Western culture is comparable, they maintain, to a drug or alcohol addiction: "Recycling is an aspirin, alleviating a rather large collective hangover . . . overconsumption." Or again, "The best way to reduce any environmental impact is not to recycle more, but to produce and dispose of less."

The tradition of issuing urgent, often moving messages to producers and consumers is rich and long-standing. But it took decades for industries themselves to really listen to them. In fact, it was not until the 1990s that leading industrialists began to recognize causes for concern. "What we thought was boundless has limits," Robert Shapiro, the chairman and chief executive officer of Monsanto, said in a 1997 interview, "and we're beginning to hit them."

The 1992 Rio Earth Summit, coinitiated by Canadian businessman Maurice Strong, was organized in response to this concern. Approximately thirty thousand people from around the world, more than a hundred world leaders, and representatives of 167 countries gathered in Rio de Janeiro to respond to troubling signals of environmental decline. To the sharp disappointment of many, no binding agreements were reached. (Strong is reported to have quipped, "There were many heads of state, but no real leaders.") But one major strategy emerged from the industrial participants: eco-efficiency. The machines of industry would be refitted with cleaner, faster, quieter engines. Industry would redeem its reputation without significantly changing its structures or compromising its drive for profit. Eco-efficiency would transform human industry from a system that takes, makes, and wastes into one that integrates economic, environmental, and ethical concerns. Industries across the globe now consider eco-efficiency to be the choice strategy of change.

What is eco-efficiency? Primarily the term means "doing more with less," a precept that has its roots in early industrialization. Henry Ford himself was adamant about lean and clean operating policies, saving his company millions of dollars by reducing waste and setting new standards with his time-saving assembly line. "You must get the most out of the power, out of the material, and out of the time," he wrote in 1926, a credo that most contemporary CEOs would proudly hang on their office walls. The linkage of efficiency with sustaining the environment was perhaps most famously articulated in *Our Common Future*, a report published in 1987 by the United Nations' World Commission on Environment and Development. *Our Common Future* warned that if pollution control was not intensified, human health, property, and ecosystems would be seriously threatened, and urban existence would become intolerable: "Industries and industrial operations should be encouraged that are more efficient in terms of resource use, that generate less pollution and waste, that are based on the use of renewable rather than non-renewable resources, and that minimize irreversible adverse impacts on human health and the environment," stated the commission in its agenda for change.

The term eco-efficiency was officially coined five years later by the Business Council for Sustainable Development, a group of forty-eight industrial sponsors including Dow, DuPont, Conagra, and Chevron, who had been asked to bring a business perspective to the Earth Summit. The council couched its call for change in practical terms, focusing on what businesses had to gain from a new ecological awareness rather than on what the environment stood to lose if industry continued current patterns. The group's report, *Changing Course*, timed for simultaneous release with the summit, stressed the importance of eco-efficiency for all companies that aimed to be competitive, sustainable, and successful in the long term. "Within a decade," predicted Stephan Schmidheiney, one of the council's founders, "it is going to be next to impossible for a business to be competitive without also being 'eco-efficient'—adding more value to a good or service while using fewer resources and releasing less pollution."

Even more quickly than Schmidheiney predicted, eco-efficiency has wended its way into industry with extraordinary success. The number of corporations adopting it continues to rise, including such big names as Monsanto, 3M (whose 3P—"Pollution Pays Program"—went into effect in 1986, before eco-efficiency was a common term), and Johnson & Johnson. The movement's famous three Rs—reduce, reuse, recycle—are steadily gaining popularity in the home as well as in the workplace. The trend stems in part from eco-efficiency's economic benefits, which can be considerable; 3M, for example, announced that by 1997 it had saved more than $750 million through pollution-prevention projects, and other companies too claim to be realizing big savings. Naturally, reducing resource consumption, energy use, emissions, and wastes has a beneficial effect on the environment as well—and on public morale. When you hear that a company like DuPont has cut its emissions of cancer-causing chemicals by almost 70 percent since 1987, you feel better. Eco-efficient industries can do something good for the environment, and people can feel less fearful about the future. Or can they?

The Four R's: Reduce, Reuse, Recycle—and Regulate

Whether it is a matter of cutting the amount of toxic waste created or emitted, or the quantity of raw materials used, or the product size itself (known in business circles as "dematerialization"), reduction is a central tenet of eco-efficiency. But reduction in any of these areas does not halt depletion and destruction—it only slows them down, allowing them to take place in smaller increments over a longer period of time.

For example, reducing the amounts of dangerous toxins and emissions released by industry is an important eco-efficient goal. It sounds unassailable, but current studies show that over time even tiny amounts of dangerous emissions can have disastrous effects on biological systems. This is a particular concern in the case of endocrine disrupters—industrial chemicals found in a variety of modern plastics and other consumer goods that appear to mimic hormones and connect with receptors in humans and other organisms. In *Our Stolen Future*, a groundbreaking report on certain synthetic chemicals and the environment, Theo Colburn, Dianne Dumanoski, and John Peterson Myers assert that "astoundingly small quantities of these hormonally active compounds can wreak all kinds of biological havoc, particularly in those exposed in the womb." Furthermore, according to these authors, many studies on the hazards of industrial chemicals have focused on cancer, while research on other kinds of damage due to exposure has only begun.

On another front, new research on particulates—microscopic particles released during incineration and combustion processes, such as those in power plants and automobiles—show that they can lodge in and damage the lungs. A 1995 Harvard study found that as many as 100,000 people die annually in the United States as a result of these tiny particles. Although regulations for controlling their release are in place, implementation does not have to begin until 2005 (and if legislation only reduces their amounts, small quantities of these particulates will still be a problem).

Another waste reduction strategy is incineration, which is often perceived as healthier than landfilling and is praised by energy efficiency proponents as "waste to energy." But waste in incinerators burns only because valuable materials, like paper and plastic, are flammable. Since these materials were never designed to be safely burned, they can release dioxins and other toxins when incinerated. In Hamburg, Germany, some trees' leaves contain such high concentrations of heavy metals from incinerator fallout that the leaves themselves must be burned, effecting a vicious cycle with a dual effect: valuable materials, such as these metals, bioaccumulate in nature to possible harmful effect and are lost to industries forever.

Air, water, and soil do not safely absorb our wastes unless the wastes themselves are completely healthy and biodegradable. Despite persistent misconceptions, even aquatic ecosystems are unable to purify and distill unsafe waste to safe levels. We have just too little knowledge about industrial pollutants and their effects on natural systems for "slowing down" to be a healthy strategy in the long term.

Finding markets to *reuse* wastes can also make industries and customers feel that something good is being done for the environment, because piles of waste appear to go "away." But in many cases these wastes—and any toxins and contaminants they contain—are simply being transferred to another place. In some developing countries, sewage sludge is recycled into animal food, but the current design and treatment of sewage by conventional sewage systems produces sludge containing chemicals that are not healthy food for any animal. Sewage sludge is also used as fertilizer, which is a well-intended attempt to make use of nutrients, but as currently processed it can contain harmful substances (like dioxins, heavy metals, endocrine disrupters, and antibiotics) that are inappropriate for fertilizing crops. Even residential sewage sludge that contains toilet paper made from recycled paper may carry dioxins. Unless materials are specifically *designed* to ultimately become safe food for nature, composting can present problems as well. When so-called biodegradable municipal wastes, including packaging and paper, are composted, the chemicals and toxins in the materials can be released into the environment. Even if these toxins exist in minute amounts, the practice may not be safe. In some cases it would actually be less dangerous to seal the materials in a landfill.

What about *recycling*? As we have noted, most recycling is actually *downcycling*; it reduces the quality of a material over time. When plastics other than those found in soda and water bottles are recycled, they are mixed with different plastics to produce a hybrid of lower quality, which is then molded into something amorphous and cheap, such as a park bench or a speed bump. Metals are often downcycled. For example, the high-quality steel used in automobiles—high-carbon, high-tensile steel—is "recycled" by melting it down with other car parts, including copper from the cables in the car, and the paint and plastic coatings. These materials lower the recycled steel's quality. More high-quality steel may be added to make the hybrid strong enough for its next use, but it will not have the material properties to make new cars again. Meanwhile the rare metals, such as copper, manganese, and chromium, and the paints, plastics, and other components that had value for industry in an unmixed, high-quality state are lost. Currently, there is no technology to separate the polymer and paint coatings from automotive metal before it is processed; therefore, even if a car were designed for disassembly, it is not technically feasible to "close the loop" for its high-quality steel. The production of one ton of copper results in the production of hundreds of tons of waste, but the copper content in some steel alloy is actually higher than it is in mined ore. Also, the presence of copper weakens steel. Imagine how useful it would be if industries had a way to recover that copper instead of constantly losing it.

Aluminum is another valuable but constantly downcycled material. The typical soda can consists of two kinds of aluminum: the walls are composed of aluminum, manganese alloy with some magnesium, plus coatings and paint, while the harder top is aluminum magnesium alloy. In conventional recycling these materials are melted together, resulting in a weaker—and less useful—product.

Lost value and lost materials are not the only concerns. Downcycling can actually increase contamination of the biosphere. The paints and plastics that are melted into recycled steel, for example, contain harmful chemicals. Electric-arc furnaces that recycle secondary steel for building materials are now a large source of dioxin emissions, an odd side effect for a supposedly environmental process. Since downcycled materials of all kinds are materially less rigorous than their predecessors, more chemicals are often added to make the materials useful again. For example, when some plastics are melted and combined, the polymers in the plastic—the chains that make it strong and flexible—shorten. Since the material properties of this recycled plastic are altered (its elasticity, clarity, and tensile strength are diminished), chemical or mineral additives may be added to attain the desired performance quality. As a result, downcycled plastic may have more additives than "virgin" plastic.

Because it was not designed with recycling in mind, paper requires extensive bleaching and other chemical processes to make it blank again for reuse. The result is a mixture of chemicals, pulp, and in some cases toxic inks that are not really appropriate for handling and use. The fibers are shorter and the paper less smooth than virgin paper, allowing an even higher proportion of particles to abrade into the air, where they can be inhaled and can irritate the nasal passages and lungs. Some people have developed allergies to newspapers, which are often made from recycled paper.

The creative use of downcycled materials for new products can be misguided, despite good intentions. For example, people may feel they are making an ecologically sound choice by buying and wearing clothing made of fibers from recycled plastic bottles. But the fibers from plastic bottles contain toxins such as antimony, catalytic residues, ultraviolet stabilizers, plasticizers, and antioxidants, which were never designed to lie next to human skin. Using downcycled paper as insulation is another current trend. But additional chemicals (such as fungicides to prevent mildew) must be added to make downcycled paper suitable for insulation, intensifying the problems already caused by toxic inks and other contaminants. The insulation might then off-gas formaldehyde and other chemicals into the home.

In all of these cases, the agenda to recycle has superseded other design considerations. Just because a material is recycled does not automatically make it ecologically benign, especially if it was not designed specifically for recycling. Blindly adopting superficial environmental approaches without fully understanding their effects can be no better—and perhaps even worse—than doing nothing.

Downcycling has one more disadvantage. It can be more expensive for businesses, partly because it tries to force materials into more lifetimes than they were originally designed for, a complicated and messy conversion and one that itself expends energy and resources. Legislation in Europe requires packaging materials that are made of aluminum and polypropylene to be recycled. But because these boxes are not designed to be recycled into new packaging (that is, to be reused by the industry to make its own product again), compliance results in additional operating costs. The components of the old packages are often downcycled into lower-quality products until they are eventually incinerated or landfilled anyway. In this instance as in many others, an ecological agenda becomes a burden for industry instead of a rewarding option.

In *Systems of Survival* the urbanist and economic thinker Jane Jacobs describes two fundamental syndromes of human civilizations: what she calls the *guardian* and *commerce*. The guardian is the government, the agency whose primary purpose is to preserve and protect the public. This syndrome is slow and serious. It reserves the right to kill—that is, it will go to war. It represents the public interest, and it is meant to shun commerce (witness conflicts over capital campaign contributions from vested interests).

Commerce, on the other hand, is the day-to-day, instant exchange of value. The name of its primary tool, currency, denotes its urgency. Commerce is quick, highly creative, inventive, constantly seeking short- and long-term advantage, and inherently honest: you can't do business with people if they aren't trustworthy. Any hybrid of these two syndromes Jacobs characterizes as so riddled with problems as to be "monstrous." Money, the tool of commerce, will corrupt the guardian. Regulation, the tool of the guardian, will slow down commerce. An example: a manufacturer might spend more money to provide an improved product under regulations, but its commercial customers, who want products quickly and cheaply, may be unwilling to absorb the extra costs. They may then find what they need elsewhere, perhaps offshore, where regulations are less stringent. In an unfortunate turnaround, the unregulated and potentially dangerous product is given a competitive edge.

For regulators who are attempting to safeguard whole industries, the readiest solutions are often those that can be applied on a very large scale, such as so-called end-of-pipe solutions, in which regulations are applied to the waste and polluting streams of a process or system. Or regulators may try to dilute or distill emissions to a more acceptable level, requiring businesses to increase ventilation or to pump more fresh air into a building because of poor indoor air quality due to offgassing materials or processes. But this "solution" to pollution—dilution—is an outdated and ineffective response that does not examine the design that caused the pollution in the first place. The essential flaw remains: badly designed materials and systems that are unsuitable for indoor use.

Jacobs sees other problems with "monstrous hybrids." Regulations force companies to comply under threat of punishment, but they seldom *reward* commerce for taking initiatives. Since regulations often require one-size-fits-all end-of-pipe solutions rather than a deeper design response, they do not directly encourage creative problem-solving. And regulation can pit environmentalists and industries against each other. Because regulations seem like a chastisement, industrialists find them annoying and burdensome. Since environmental goals are typically forced upon business by the guardian—or are simply perceived as an added dimension outside crucial operating methods and goals—industrialists see environmental initiatives as inherently uneconomic.

We do not mean to lambaste those who are working with good intentions to create and enforce laws meant to protect the public good. In a world where designs are unintelligent and destructive, regulations can reduce immediate deleterious effects. But ultimately a regulation is a signal of design failure. In fact, it is what we call a *license to harm*: a permit issued by a government to an industry so that it may dispense sickness, destruction, and death at an "acceptable" rate. But as we shall see, good design can require no regulation at all.

Eco-efficiency is an outwardly admirable, even noble, concept, but it is not a strategy for success over the long term, because it does not reach deep enough. It works within the same system that caused the problem in the first place, merely slowing it down with moral proscriptions and punitive measures. It presents little more than an illusion of change. Relying on eco-efficiency to save the environment will in fact achieve the opposite; it will let industry finish off everything, quietly, persistently, and completely.

Remember the retroactive design assignment that we applied to the Industrial Revolution in Chapter One? If we were to take a similar look at industry under the influence of the eco-efficiency movement, the results might look like this:

Design a system of industry that will:

- release *fewer* pounds of toxic wastes into the air, soil, and water every year
- measure prosperity by *less* activity
- *meet* the stipulations of thousands of complex regulations to keep people and natural systems from being poisoned too quickly
- produce *fewer* materials that are so dangerous that they will require future generations to maintain constant vigilance while living in terror
- result in *smaller* amounts of useless waste
- put *smaller* amounts of valuable materials in holes all over the planet, where they can never be retrieved.

Plainly put, eco-efficiency only works to make the old, destructive system a bit less so. In some cases, it can be more pernicious, because its workings are more subtle and long-term. An ecosystem might actually have more of a chance to become healthy and whole again after a quick collapse that leaves some niches intact than with a slow, deliberate, and efficient destruction of the whole.

Efficient—at What?

As we have seen, even before the term *eco-efficiency* was coined, industry generally viewed efficiency as a virtue. We would like to question the general goal of efficiency for a system that is largely destructive.

Consider energy-efficient buildings. Twenty years ago in Germany, the standard rate of oil use for heating and cooling the average house was 30 liters per square meter per year. Today, with high-efficiency housing, that number has plummeted to 1.5 liters of oil per square meter. Increased efficiency is often achieved through better insulation (such as plastic coatings in potential air-exchange areas so that less air comes into the building from outside) and smaller, leak-proof windows. These strategies are meant to optimize the system and reduce wasted energy. But by reducing air-exchange rates, efficient homeowners are actually strengthening the concentration of indoor air pollution from poorly designed materials and products in the home. If indoor air quality is poor because of crude products and building materials, then people require more fresh air to circulate throughout the building, not less.

Overly efficient buildings can also be dangerous. Several decades ago the Turkish government created inexpensive housing by designing and constructing apartments and houses which were built "efficiently," with a minimum of steel and concrete. During the 1999 earthquakes, however, this housing easily collapsed, while older, "inefficient" buildings held up better. In the short term, people saved money on housing, but in the long term, the efficiency strategy turned out to be dangerous. What social benefit does cheap, efficient housing provide if it also exposes people to more dangers than traditional housing?

Efficient agriculture can perniciously deplete local landscapes and wildlife. The contrast between the former East Germany and West Germany is a good example. Traditionally, the average amount of wheat produced in eastern Germany per acre has been only half that of western Germany, because the agricultural industry in the west is more modern and efficient. The eastern region's "inefficient," more old-fashioned agriculture is actually better for environmental health: it has larger wetland areas that have not been drained and overtaken by monocultural crops, and they contain more rare species—for example, three thousand nesting pairs of storks, compared with 240 pairs in the more developed western lands. These wild marshes and wetland areas provide vital centers for breeding, nutrient cycling, and water absorption and purification. Today agriculture all over Germany is becoming more efficient, destroying wetlands and other habitats, resulting in rising extinction rates.

Eco-efficient factories are held up as models of modern manufacturing. But in truth many of them are only distributing their pollution in less obvious ways. Less efficient factories, instead of sending emissions through high smokestacks into other areas far from the site (or importing them), tend to contaminate local areas. At least local destruction tends to be more visible and comprehensible: if you know what you are dealing with, you may be horrified enough to do something about it. Efficient destruction is harder to detect and thus harder to stop.

In a philosophical sense, efficiency has no independent value: it depends on the value of the larger system of which it is a part. An efficient Nazi, for example, is a terrifying thing. If the aims are questionable, efficiency may even make destruction more insidious.

Last but not least, efficiency isn't much fun. In a world dominated by efficiency, each development would serve only narrow and practical purposes. Beauty, creativity, fantasy, enjoyment, inspiration, and poetry would fall by the wayside, creating an unappealing world indeed. Imagine a fully efficient world: an Italian dinner would be a red pill and a glass of water with an artificial aroma. Mozart would hit the piano with a two-by-four. Van Gogh would use one color. Whitman's sprawling "Song of Myself" would fit on a single page. And what about efficient sex? An efficient world is not one we envision as delightful. In contrast to nature, it is downright parsimonious.

This is not to condemn *all* efficiency. When implemented as a tool within a larger, effective system that intends overall positive effects on a wide range of issues—not simply economic ones—efficiency can actually be valuable. It is valuable too when conceived as a transitional strategy to help current systems slow down and turn around. But as long as modern industry is so destructive, attempting only to make it less bad is a fatally limited goal.

The "be less bad" environmental approaches to industry

have been crucial in sending important messages of environmental concern—messages that continue to catch the public's attention and to spur important research. At the same time, they forward conclusions that are less useful. Instead of presenting an inspiring and exciting vision of change, conventional environmental approaches focus on what *not* to do. Such proscriptions can be seen as a kind of guilt management for our collective sins, a familiar placebo in Western culture.

In very early societies, repentance, atonement, and sacrifice were typical reactions to complex systems, like nature, over which people felt they had little control. Societies around the world developed belief systems based on myth in which bad weather, famine, or disease meant one had displeased the gods, and sacrifices were a way to appease them. In some cultures, even today, one must sacrifice something of value in order to regain the blessing of the gods (or god) and reestablish stability and harmony.

Environmental destruction is a complex system in its own right—widespread, with deeper causes that are difficult to see and understand. Like our ancestors, we may react automatically, with terror and guilt, and we may look for ways to purge ourselves—which the "eco-efficiency" movement provides in abundance, with its exhortations to consume and produce less by minimizing, avoiding, reducing, and sacrificing. Humans are condemned as the one species on the planet guilty of burdening it beyond what it can withstand; as such, we must shrink our presence, our systems, our activities, and even our population so as to become almost invisible. (Those who believe population is the root of our ills think people should mostly stop having children.) The goal is zero: zero waste, zero emissions, zero "ecological footprint."

As long as human beings are regarded as "bad," zero is a good goal. But to be less bad is to accept things as they are, to believe that poorly designed, dishonorable, destructive systems are the *best* humans can do. This is the ultimate failure of the "be less bad" approach: a failure of the imagination. From our perspective, this is a depressing vision of our species' role in the world.

What about an entirely different model? What would it mean to be 100 percent good?

Chapter Three

Eco-Effectiveness

Here's a tale of three books.

The first is familiar. It is about five inches by eight, compact and pleasant to hold. Dark ink makes a crisp impression on the creamy paper. It has a colorful jacket and a sturdy cardboard cover. In many respects, it is an intelligently conceived object, designed—as were its very similar predecessors, hundreds of years ago—with portability and durability in mind. Hundreds of users may check it out of the library. They take it to bed, on the train, to the beach.

Yet attractive, functional, and durable as it is, the book will not last forever—nor, if it is "beach reading," do we necessarily expect it to. What happens when it is discarded? The paper came from trees, so natural diversity and soils have already been depleted to keep us in reading matter. Paper is biodegradable, but the inks that printed so crisply on the paper and created the striking image on the jacket contain carbon black and heavy metals. The jacket is not really paper, but an amalgam of materials—wood pulp, polymers, and coatings, as well as inks, heavy metals, and halogenated hydrocarbons. It cannot be safely composted, and if it is burned, it produces dioxins, some of the most dangerous cancer-causing material ever created by humans.

Enter book number two. It too is rather familiar to contemporary eyes. It has the usual book shape and format, but the paper—a dull beige—is thin and porous. It has no jacket, and the cover, like the inside, is printed in a single shade of ink. It may seem a little drab, but it has a humble, "earth-friendly" look that is instantly recognizable to the environmentally minded. And indeed the book is the product of a concerted attempt to be eco-efficient. It is printed on recycled paper—hence the beige—with soy-based inks. In addition, its designers strove to "dematerialize," to use less of everything; witness the thin, uncoated text stock and the absence of a jacket. Unfortunately, the ink shows through the flimsy paper, and the lack of contrast between ink and page strains the eyes. The skimpy binding is a little weak to boot. The book isn't exactly reader-friendly—good thing it's eco-friendly.

Or is it?

Its designers thought long and hard about what kind of paper to use; every choice had drawbacks. Initially they thought chlorine-free paper might be a good way to go, because they knew that chlorine presents a serious problem for ecosystems and human health (by creating dioxins, for example). But they discovered that totally chlorine-free paper required virgin pulp, because any recycled paper in the mix would already have been bleached. In fact, paper made from any kind of wood pulp probably contains some chlorine, because chlorinated salt occurs naturally in trees. What a quandary: pollute rivers or chew up forests. They ended up choosing paper with the greatest recycled content, avoiding what to their minds would be a greater offense. Soy-based inks posed another dilemma, because they might include halogenated hydrocarbons or other toxins that become more bioavailable in these water-soluble eco-friendly inks than they would be in conventional solvent-based inks. For acceptable durability, the cover was coated, so it isn't recyclable with the rest of the book, and because of its already high recycled content, the paper's fibers have about reached the limits of further use. Once again, being less bad proves to be a fairly unappealing option, practically, aesthetically, and environmentally.

Imagine if we were to rethink the entire concept of a book, considering not only the practicalities of manufacture and use but the pleasures that might be brought to both. Enter book three, the book of the future.

Is it an electronic book? Perhaps—that form is still in its infancy. Or perhaps it takes another form as yet unimagined by us. But many people find the form of the traditional book both convenient and delightful. What if we reconceived not the shape of the object but the materials of which it is made, in the context of its relationship to the natural world? How could it be a boon to both people and the environment?

We might begin by considering whether paper itself is a proper vehicle for reading matter. Is it fitting to write our history on the skin of fish with the blood of bears, to echo writer Margaret Atwood? Let's imagine a book that is not a tree. It is not even paper. Instead, it is made of plastics developed around a completely different paradigm for materials, polymers that are infinitely recyclable at the same level of quality—that have been designed with their future life foremost in mind, rather than as an awkward afterthought. This "paper" doesn't require cutting down trees or leaching chlorine into waterways. The inks are nontoxic and can be washed off the polymer with a simple and safe chemical process or an extremely hot water bath, from either of which they can be recovered and reused. The cover is made from a heavier grade of the same polymer as the rest of the book, and the glues are made of compatible ingredients, so that once the materials are no longer needed in their present form, the entire book can be reclaimed by the publishing industry in a simple one-step recycling process.

Nor is the reader's pleasure and convenience an afterthought to environmentally responsible design. The pages are white and have a sensuous smoothness, and unlike recycled paper, they will not yellow with age. The ink won't rub off on the reader's fingers. Although its next life has already been imagined, this book is durable enough to last for many generations. It's even waterproof, so you can read it at the beach, even in the hot tub. You'd buy it, carry it, and read it not as a badge of austerity—and not only for its content—but for its sheer tactile pleasure. It celebrates its materials rather than apologizing for them. Books become books become books over and over again, each incarnation a sparkling new vehicle for fresh images and ideas. Form follows not just function but the evolution of the medium itself, in the endlessly propagating spirit of the printed word.

The assignment that leads to the design of this third book is to tell a story within the very molecules of its pages. Not the old tale of damage and despair, but one of abundance and renewal, human creativity and possibility. And although the book you hold in your hands is not yet that book, it is a step in that direction, a beginning to the story.

We did not design the materials of this book. After years of analyzing and testing polymers to replace paper, we were delighted when designer Janine James happened to mention our search to Charlie Melcher of Melcher Media. Melcher was working with a paper adapted from a polymer blend that had been used to label detergent bottles, so that the labels could be recycled along with the bottles instead of being burned off. For "selfish" reasons, they wanted an alternative to the usual "monstrous hybrid." Charlie was in search of a waterproof paper on which he could print books that could be read in the bath or at the beach. He knew its qualities extended beyond imperviousness to water and was eager to have us explore its eco-effective promise. When Michael tested it, he found that it off-gassed similarly to a conventional book. But it could be recycled, and more to the point, it has the potential to be *up*cycled: dissolved and remade as polymer of high quality and usefulness.

Once we set about designing with such missions in mind—the short-term usefulness, convenience, and aesthetic pleasure of the product together with the ongoing life of its materials—the process of innovation begins in earnest. We leave aside the old model of product-and-waste, and its dour offspring, "efficiency," and embrace the challenge of being not efficient but *effective* with respect to a rich mix of considerations and desires.

Consider the Cherry Tree

Consider the cherry tree: thousands of blossoms create fruit for birds, humans, and other animals, in order that one pit might eventually fall onto the ground, take root, and grow. Who would look at the ground littered with cherry blossoms and complain, "How inefficient and wasteful!" The tree makes copious blossoms and fruit without depleting its environment. Once they fall on the ground, their materials decompose and break down into nutrients that nourish microorganisms, insects, plants, animals, and soil. Although the tree actually makes more of its "product" than it needs for its own success in an ecosystem, this abundance has evolved (through millions of years of success and failure or, in business terms, R&D), to serve rich and varied purposes. In fact, the tree's fecundity nourishes just about everything around it.

What might the human built world look like if a cherry tree had produced it?

We know what an eco-efficient building looks like. It is a big energy saver. It minimizes air infiltration by sealing places that might leak. (The windows do not open.) It lowers solar income with dark-tinted glass, diminishing the cooling load on the building's air-conditioning system and thereby cutting the amount of fossil-fuel energy used. The power plant in turn releases a smaller amount of pollutants into the environment, and whoever foots the electric bill spends less money. The local utility honors the building as the most energy-saving in its area and holds it up as a model for environmentally conscious design. If all buildings were designed and built this way, it proclaims, businesses could do right by the environment and save money at the same time.

Here's how we imagine the cherry tree would do it: during the daytime, light pours in. Views of the outdoors through large, untinted windows are plentiful—each of the occupants has five views from wherever he or she happens to sit. Delicious, affordable food and beverages are available to employees in a café that opens onto a sun-filled courtyard. In the office space, each of them controls the flow of fresh air and the temperature of their personal breathing zones. The windows open. The cooling system maximizes natural airflows, as in a hacienda: at night, the system flushes the building with cool evening air, bringing the temperature down and clearing the rooms of stale air and toxins. A layer of native grasses covers the building's roof, making it more attractive to songbirds and absorbing water runoff, while at the same time protecting the roof from thermal shock and ultraviolet degradation.

In fact, this building is just as energy-efficient as the first, but that is a side effect of a broader and more complex design goal: to create a building that celebrates a range of cultural and natural pleasures—sun, light, air, nature, even food—in order to enhance the lives of the people who work there. During construction, certain elements of the second building did cost a little more. For example, windows that open are more expensive than windows that do not. But the nighttime cooling strategy cuts down on the need for air-conditioning during the day. Abundant daylight diminishes the need for fluorescent light. Fresh air makes the indoor spaces more pleasurable, a perk for current employees and a lure to potential ones—and thus an effect with economic as well as aesthetic consequences. (Securing and supporting a talented and productive workforce is one of a CFO's primary goals, because the carrying cost of people—recruiting, employing, and retaining them—is a hundred times as great as the carrying cost of the average building.) In its every element, the building expresses the client's and architects' vision of a life-centered community and environment. We know, because Bill's firm led the team that designed it.

We brought the same sensibility to designing a factory for Herman Miller, the office-furniture manufacturer. We wanted to give workers the feeling that they'd spent the day outdoors, unlike workers in the conventional factory of the Industrial Revolution, who might not see daylight until the weekend. The offices and manufacturing space that we designed for Herman Miller were built for only 10 percent more money than it would have cost to erect a standard prefabricated metal factory building. We designed the factory around a tree-lined interior conceived as a brightly daylit "street" that ran the entire length of the building. There are rooftop skylights everywhere the workers are stationed, and the manufacturing space offers views of both the internal street and the outdoors, so that even as they work indoors, employees get to participate in the cycles of the day and the seasons. (Even the truck docks have windows.) The factory was designed to celebrate the local landscape and to invite indigenous species back to the site instead of scaring them away. Storm water and waste water are channeled through a series of connected wetlands that clean them, in the process lightening the load on the local river, which already suffers serious flooding because of runoff from roofs, parking lots, and other impervious surfaces.

An analysis of the factory's dramatic productivity gains has shown that one factor was "biophilia"—people's love of the outdoors. Retention rates have been impressive. A number of workers who left for higher wages at a competitor's factory returned in a few weeks. When asked why, they told the management they couldn't work "in the dark." They were young people who had entered the workforce only recently and had never worked in a "normal" factory before.

These buildings represent only the beginnings of eco-effective design; they do not yet exemplify, in every way, the principles we espouse. But you might start to envision the difference between eco-efficiency and eco-effectiveness as the difference between an airless, fluorescent-lit gray cubicle and a sunlit area full of fresh air, natural views, and pleasant places to work, eat, and converse.

Peter Drucker has pointed out that it is a manager's job to "do things right." It is an executive's job to make sure "the right things" get done. Even the most rigorous eco-efficient business paradigm does not challenge basic practices and methods: a shoe, building, factory, car, or shampoo can remain fundamentally ill-designed even as the materials and processes involved in its manufacture become more "efficient." Our concept of eco-effectiveness means working on the right things—on the right products and services and systems—instead of making the wrong things less bad. Once you are doing the right things, then doing them "right," with the help of efficiency among other tools, makes perfect sense.

If nature adhered to the human model of efficiency, there would be fewer cherry blossoms, and fewer nutrients. Fewer trees, less oxygen, and less clean water. Fewer songbirds. Less diversity, less creativity and delight. The idea of nature being more efficient, dematerializing, or even not "littering" (imagine zero waste or zero emissions for nature!) is preposterous. The marvelous thing about effective systems is that one wants more of them, not less.

What Is Growth?

Ask a child about growth, and she will probably tell you it is a good thing, a natural thing—it means getting bigger, healthier, and stronger. The growth of nature (and of children) is usually perceived as beautiful and healthy. Industrial growth, on the other hand, has been called into question by environmentalists and others concerned about the rapacious use of resources and the disintegration of culture and environment. Urban and industrial growth is often referred to as a cancer, a thing that grows for its own sake and not for the sake of the organism it inhabits. (As Edward Abbey wrote, "Growth for growth's sake is a cancerous madness.")

Conflicting views of growth were a recurrent source of tension on President Clinton's original Council on Sustainable Development, a group of twenty-five representatives of business, government, diverse social groups, and environmental organizations that met from 1993 to 1999. The commercial members' belief that commerce is inherently required to perpetuate itself, that it must seek growth in order to fuel its continued existence, brought them to loggerheads with the environmentalists, to whom commercial growth meant more sprawl, more loss of ancient forests, wild places, and species, and more pollution, toxification, and global warming. Their desire for a no-growth scenario naturally frustrated the commercial players, for whom "no growth" could have only negative consequences. The perceived conflict between nature and industry made it look as if the values of one system must be sacrificed to the other.

But unquestionably there are things we all want to grow, and things we don't want to grow. We wish to grow education and not ignorance, health and not sickness, prosperity and not destitution, clean water and not poisoned water. We wish to improve the quality of life.

The key is not to make human industries and systems smaller, as efficiency advocates propound, but to design them to get bigger and better in a way that replenishes, restores, and nourishes the rest of the world. Thus the "right things" for manufacturers and industrialists to do are those that lead to good growth—more niches, health, nourishment, diversity, intelligence, and abundance—for this generation of inhabitants on the planet and for generations to come.

Let's take a closer look at that cherry tree.

As it grows, it seeks its own regenerative abundance. But this process is not single-purpose. In fact, the tree's growth sets in motion a number of positive effects. It provides food for animals, insects, and microorganisms. It enriches the ecosystem, sequestering carbon, producing oxygen, cleaning air and water, and creating and stabilizing soil. Among its roots and branches and on its leaves, it harbors a diverse array of flora and fauna, all of which depend on it and on one another for the functions and flows that support life. And when the tree dies, it returns to the soil, releasing, as it decomposes, minerals that will fuel healthy new growth in the same place.

The tree is not an isolated entity cut off from the systems around it: it is inextricably and productively engaged with them. This is a key difference between the growth of industrial systems as they now stand and the growth of nature.

Consider a community of ants. As part of their daily activity, they:

- safely and effectively handle their own material wastes and those of other species
- grow and harvest their own food while nurturing the ecosystem of which they are a part
- construct houses, farms, dumps, cemeteries, living quarters, and food-storage facilities from materials that can be truly recycled
- create disinfectants and medicines that are healthy, safe, and biodegradable
- maintain soil health for the entire planet.

Individually we are much larger than ants, but collectively their biomass exceeds ours. Just as there is almost no corner of the globe untouched by human presence, there is almost no land habitat, from harsh desert to inner city, untouched by some species of ant. They are a good example of a population whose density and productiveness are not a problem for the rest of the world, because everything they make and use returns to the cradle-to-cradle cycles of nature. All their materials, even their most deadly chemical weapons, are biodegradable, and when they return to the soil, they supply nutrients, restoring in the process some of those that were taken to support the colony. Ants also recycle the wastes of other species; leaf-cutter ants, for example, collect decomposing matter from the Earth's surface, carry it down into their colonies, and use it to feed the fungus gardens that they grow underground for food. During their movements and activities, they transport minerals to upper layers of soil, where plant life and fungi can use them as nutrients. They turn and aerate the soil and make passageways for water drainage, playing a vital role in maintaining soil fecundity and health. They truly are, as biologist E. O. Wilson has pointed out, the little things that run the world. But although they may run the world, they do not *overrun* it. Like the cherry tree, they make the world a better place.

Some people use the term *nature's services* to refer to the processes by which, without human help, water and air are purified; erosion, floods, and drought are mitigated; materials are detoxified and decomposed; soil is created and its fertility renewed; ecological equilibrium and diversity are maintained; climate is stabilized; and, not least, aesthetic and spiritual satisfaction is provided to us. We don't like this focus on *services*, since nature does not do any of these things just to serve people. But it is useful to think of these processes as part of a dynamic interdependence, in which many different organisms and systems support one another in multiple ways. The consequences of growth—increases in insects, microorganisms, birds, water cycling, and nutrient flows—tend toward the positive kind that enrich the vitality of the whole ecosystem. The consequences of a new strip mall, on the other hand, while they may have some immediate local benefits (jobs, more money circulating through the local economy) and may even boost the country's overall GDP, are gained at the expense of a decline in overall quality of life—increased traffic, asphalt, pollution, and waste—that ultimately undermines even some of the mall's ostensible benefits.

Typically, conventional manufacturing operations have predominantly negative side effects. In a textile factory, for example, water may come in clean, but it goes out contaminated with fabric dyes, which usually contain toxins such as cobalt, zirconium, other heavy metals, and finishing chemicals. Solid wastes from fabric trimmings and loom clippings present another problem, as much of the material used for textiles is petrochemical-based. Effluents and sludge from production processes cannot be safely deposited into ecosystems, so they are often buried or burned as hazardous waste. The fabric itself is sold all over the world, used, then thrown "away"—which usually means it is either incinerated, releasing toxins, or placed in a landfill. Even in the rather short life span of the fabric, its particles have abraded into the air and been taken into people's lungs. All this in the name of efficient production.

Just about every process has side effects. But they can be deliberate and sustaining instead of unintended and pernicious. We can be humbled by the complexity and intelligence of nature's activity, and we can also be inspired by it to design some positive side effects to our own enterprises instead of focusing exclusively on a single end.

Eco-effective designers expand their vision from the primary purpose of a product or system and consider the whole. What are its goals and potential effects, both immediate and wide-ranging, with respect to both time and place? What is the entire system—cultural, commercial, ecological—of which this made thing, and way of making things, will be a part?

Once upon a Roof

Once you begin to consider the larger picture, the most familiar features of human fabrication begin to shape-shift. An ordinary roof is a good example. Conventional roofing surfaces are infamously among the most expensive parts of a building to maintain: baking under the sun all day, they are exposed to relentless ultraviolet degradation, and dramatic variations between daytime and nighttime temperatures subject them to constant thermal shock. But in the larger context, they reveal themselves as part of the growing landscape of impervious surfaces (along with paved roads, parking lots, sidewalks, and buildings themselves) that contribute to flooding, heat up cities in the summertime (dark surfaces absorb and re-emit solar energy), and deplete habitat for many species.

If we viewed these effects piecemeal, we might attempt to address the flooding problem by calling for regulations requiring big retention ponds for storm water. We'd "solve" the heat problem by providing additional air-conditioning units to buildings in the area, doing our best to ignore the fact that the new units would contribute to the higher ambient temperatures that made them necessary in the first place. As for shrinking habitat, well, we'd likely throw up our hands. Isn't wildlife an inevitable casualty of urban growth?

We have been working with a kind of roofing that responds to all of these issues, including the economic ones. It is a light layer of soil, a growing matrix, covered with plants. It maintains the roof at a stable temperature, providing free evaporative cooling in hot weather and insulation in cold weather, and shields it from the sun's destructive rays, making it last longer. In addition, it makes oxygen, sequesters carbon, captures particulates like soot, and absorbs storm water. And that's not all: it looks far more attractive than naked asphalt and, with the storm-water management, saves money that would be lost to regulatory fees and flood damage. In appropriate locales, it can even be engineered to produce solar-generated electricity.

If this sounds like a novel idea, it's not. It is based on centuries-old building techniques. (In Iceland, for example, many old farms were built with stones, wood, and sod, and grass for roofs.) And it is widely used in Europe, where tens of millions of square feet of such roofing already exist. Enhanced by today's sophisticated technology and engineering, this approach to roofing is effective on multiple levels, not least of which is its ability to capture the public imagination. We helped Mayor Richard Daley put a garden on the roof of Chicago's city hall, and he foresees a whole city covered with green roofs that will not only keep it cool but produce solar energy and grow food and flowers, as well as providing soothing green sanctuary from busy urban streets to birds and people alike.

Beyond Control

Taking an eco-effective approach to design might result in an innovation so extreme that it resembles nothing we know, or it might merely show us how to optimize a system already in place. It's not the solution itself that is necessarily radical but the shift in perspective with which we begin, from the old view of nature as something to be controlled to a stance of engagement.

For thousands of years, people struggled to maintain the boundaries between human and natural forces; to do so was often necessary to their survival. Western civilization in particular has been shaped by the belief that it is the right and duty of human beings to shape nature to better ends; as Francis Bacon put it, "Nature being known, it may be master'd, managed, and used in the services of human life."

Today few natural disasters can really threaten those of us in the industrialized nations. On a day-to-day basis, we are fairly safe from all but the most serious epidemics and climatic events: earthquakes, hurricanes, volcanoes, floods, plagues, perhaps a meteor. Yet we still cling to a mental model of civilization based on the practices of our ancestors, who hacked and plowed their way through a difficult wilderness. Overwhelming and controlling nature is not only the reigning trend, it has even become an aesthetic preference. The hedges or borders of the modern lawn sharply distinguish what is "natural" from what is "civilized." In a city landscape of asphalt, concrete, steel, and glass, nature's excess may be considered messy, even useless, something to be limited to a few carefully sculpted gardens and trees. What autumn leaves there are must be quickly gathered from the ground, placed in plastic bags, and landfilled or burned rather than composted. Instead of trying to optimize nature's abundance, we automatically try to get it out of the way. For many of us used to a culture of control, nature in its untamed state is neither a familiar nor a welcoming place.

To emphasize this point, Michael likes to tell the story of the forbidden cherry tree. In 1986 several people in a neighborhood in Hannover, Germany, decided they wanted to plant a cherry tree on their street. They thought such an addition would provide habitat for songbirds and pleasure for people who might want to eat the cherries, pluck a blossom or two, or simply admire the tree's beauty. It seemed an easy enough decision, with only positive effects. But the tree was not so easily transposed from their imaginations to real life. According to zoning laws in that neighborhood, a new cherry-tree planting would not be legal. What the residents viewed as delightful, the legislature viewed as a risk. People might slip on fallen cherries and cherry blossoms. Fruit trees with dangling fruit might lure children to climb them—a liability if a child fell and got hurt. The cherry tree was simply not efficient enough for the legislators: it was messy, creative, unpredictable. It could not be controlled or anticipated. The system was not set up to handle something of that kind. The neighbors pressed on, however, and eventually they were granted special permission to plant the tree.

The forbidden fruit tree is a useful metaphor for a culture of control, for the barriers erected and maintained—

whether physical or ideological—between nature and human industry. Sweeping away, shutting out, and controlling nature's imperfect abundance are implicit features of modern design, ones rarely if ever questioned. *If brute force doesn't work, you're not using enough of it.*

As we know from our own work, paradigms sometimes shift not only because of new ideas but because of evolving tastes and trends. Contemporary preferences are already tending toward greater diversity. Michael tells another story: in 1982 his mother's garden, which was full of vegetables, herbs, wildflowers, and many other strange and wonderful plants, was determined by town legislators to be too messy, too "wild." She was asked to pay a fine. Rather than bow down to this "minimization demand," as Michael calls it, she decided to continue growing the kind of garden she loved and to pay a yearly fine for the right to do so. Ten years later this very same garden won a local award for creating habitat for songbirds. What had changed? The public taste, the prevailing aesthetic. It is now fashionable to grow a garden that looks "wild."

Imagine the fruits of such a shift on a large scale.

Becoming a Native

There is some talk in science and popular culture about colonizing other planets, such as Mars or the moon. Part of this is just human nature: we are curious, exploring creatures. The idea of taming a new frontier has a compelling, even romantic, pull, like that of the moon itself. But the idea also provides rationalization for destruction, an expression of our hope that we'll find a way to save ourselves if we trash our planet. To this speculation, we would respond: If you want the Mars experience, go to Chile and live in a typical copper mine. There are no animals, the landscape is hostile to humans, and it would be a tremendous challenge. Or, for a moonlike effect, go to the nickel mines of Ontario.

Seriously, humans evolved on the Earth, and we are meant to be here. Its atmosphere, its nutrients, its natural cycles, and our own biological systems evolved together and support us here, now. Humans were simply not designed by evolution for lunar conditions. So while we recognize the great scientific value of space exploration and the exciting potential of new discovery there, and while we applaud technological innovations that enable humans to "boldly go where no man has gone before," we caution: Let's not make a big mess here and go somewhere less hospitable even if we figure out how. Let's use our ingenuity to stay here; to become, once again, native to this planet.

This affirmation does not mean that we advocate returning to a pretechnological state. We believe that humans can incorporate the best of technology and culture so that our civilized places reflect a new view. Buildings, systems, neighborhoods, and even whole cities can be entwined with surrounding ecosystems in ways that are mutually enriching. We agree that it is important to leave some natural places to thrive on their own, without undue human interference or habitation. But we also believe that industry can be so safe, effective, enriching, and intelligent that it need not be fenced off from other human activity. (This could stand the concept of zoning on its head;

when manufacturing is no longer dangerous, commercial and residential sites can exist alongside factories, to their mutual benefit and delight.)

The Menominee tribe of Wisconsin, wood harvesters for many generations, use a logging method that lets them profit from nature while allowing it to thrive. Conventional logging operations are focused on producing a certain amount of carbohydrate (wood pulp) for use. This agenda is single-purpose and utilitarian: it does not count how many species of birds the forest may harbor, or how its slopes stay stable, or what occasions for recreation and respite—as well as resources—it provides and could continue to provide to future generations. The Menominee often cut only the weaker trees, leaving the strong mother trees and enough of the upper canopy for squirrels and other arboreal animals to continuously inhabit. This strategy has been enormously productive; it has allowed the forest to thrive while supplying the tribe with commercial resources. In 1870 the Menominee counted 1.3 billion standing board feet of timber—what in the timber industry is tellingly known as "stumpage"—on a 235,000-acre reservation. Over the years they have harvested 2.25 billion feet, yet today they have 1.7 billion standing feet—a slight increase. One might say they have figured out what the forest can productively offer them instead of considering only what they want. (It's important to note here that this particular form of forestry is not necessarily universal in its potential applications. In some instances—including restorative work, in which you might remove a monocultural forest to plant a more diverse system—clear-cutting appears to be a successful management tool. As the

Forest Stewardship Council notes, there are no absolutes about method.)

Kai Lee, a professor of environmental science at Williams College, tells an enlightening story about native peoples' view of place. In 1986 Lee was involved in plans for the long-term storage of radioactive wastes at the Hanford Reservation, a large site in central Washington State, where the United States government had produced plutonium for nuclear weapons. He spent a morning with scientists discussing how to mark a waste site so that even in the distant future, people would not accidentally drill for water there or otherwise bring about harmful exposures and releases. During a break he saw several members of the Yakima Indian Nation, whose traditional lands include much of the Hanford Reservation. They had come there to talk with federal officials about another matter. The Yakima were surprised—even amused—at Kai's concern over their descendants' safety. "Don't worry," they assured him. "We'll tell them where it is." As Kai pointed out to us, "Their conception of themselves and their place was not historical, as mine was, but eternal. This would always be their land. They would warn others not to mess with the wastes we'd left."

We are not leaving this land either, and we will begin to become native to it when we recognize this fact.

The New Design Assignment

An old joke about efficiency: An olive-oil vendor returns from the marketplace and complains to a friend, "I can't make money selling olive oil! By the time I feed the donkey that carries my oil to market, most of my profit is gone." His friend suggests he feed the donkey a little less. Six weeks later they meet again at the marketplace. The oil seller is in poor shape, with neither money nor donkey. When his friend asks what happened, the vendor replies, "Well, I did as you said. I fed the donkey a little less, and I began to do really well. So I fed him even less, and I did even better. But just at the point where I was becoming really successful, he died!"

Is our goal to starve ourselves? To deprive ourselves of our own culture, our own industries, our own presence on the planet, to aim for zero? How inspiring a goal is that? Wouldn't it be wonderful if, rather than bemoaning human industry, we had reason to champion it? If environmentalists as well as automobile makers could applaud every time someone exchanged an old car for a new one, because new cars purified the air and produced drinking water? If new buildings imitated trees, providing shade, songbird habitat, food, energy, and clean water? If each new addition to a human community deepened ecological and cultural as well as economic wealth? If modern societies were perceived as increasing assets and delights on a very large scale, instead of bringing the planet to the brink of disaster?

We would like to suggest a new design assignment. Instead of fine-tuning the existing destructive framework, why don't people and industries set out to create the following:

- buildings that, like trees, produce more energy than they consume and purify their own waste water

- factories that produce effluents that are drinking water
- products that, when their useful life is over, do not become useless waste but can be tossed onto the ground to decompose and become food for plants and animals and nutrients for soil; or, alternately, that can return to industrial cycles to supply high-quality raw materials for new products
- billions, even trillions, of dollars' worth of materials accrued for human and natural purposes each year
- transportation that improves the quality of life while delivering goods and services
- a world of abundance, not one of limits, pollution, and waste.

Chapter Four

Waste Equals Food

Nature operates according to a system of nutrients and metabolisms in which there is no such thing as waste. A cherry tree makes many blossoms and fruit to (perhaps) germinate and grow. That is why the tree blooms. But the extra blossoms are far from useless. They fall to the ground, decompose, feed various organisms and microorganisms, and enrich the soil. Around the world, animals and humans exhale carbon dioxide, which plants take in and use for their own growth. Nitrogen from wastes is transformed into protein by microorganisms, animals, and plants. Horses eat grass and produce dung, which provides both nest and nourishment for the larvae of flies. The Earth's major nutrients—carbon, hydrogen, oxygen, nitrogen—are cycled and recycled. Waste equals food.

This cyclical, cradle-to-cradle biological system has nourished a planet of thriving, diverse abundance for millions of years. Until very recently in the Earth's history, it was the only system, and every living thing on the planet belonged to it. Growth was good. It meant more trees, more species, greater diversity, and more complex, resilient ecosystems. Then came industry, which altered the natural equilibrium of materials on the planet. Humans took substances from the Earth's crust and concentrated, altered, and synthesized them into vast quantities of material that cannot safely be returned to soil. Now material flows can be divided into two categories: biological mass and technical—that is, industrial—mass.

From our perspective, these two kinds of material flows on the planet are just *biological and technical nutrients*. Biological nutrients are useful to the biosphere, while technical nutrients are useful for what we call the *technosphere*, the systems of industrial processes. Yet somehow we have evolved an industrial infrastructure that ignores the existence of nutrients of either kind.

From Cradle-to-Cradle to Cradle-to-Grave: A Brief History of Nutrient Flows

Long before the rise of agriculture, nomadic cultures wandered from place to place searching for food. They needed to travel light, so their possessions were few—some jewelry and a few tools, bags or clothes made of animal skins, baskets for roots and seeds. Assembled from local materials, these things, when their use was over, could easily decompose and be "consumed" by nature. The more durable objects, such as weapons of stone and flint, might be discarded. Sanitation was not a problem because the nomads were constantly moving. They could leave their biological wastes behind to replenish soil. For these people, there truly was an "away."

Early agricultural communities continued to return biological wastes to the soil, replacing nutrients. Farmers rotated crops, letting fields lie fallow in turn until nature made them fertile again. Over time new agricultural tools and techniques led to quicker food production. Populations swelled, and many communities began to take more resources and nutrients than could be naturally restored. With people more tightly packed, sanitation became a problem. Societies began to find ways to get rid of their wastes. They also began to take more and more nutrients from the soil and to eat up resources (such as trees) without replacing them at an equal rate.

There is an old Roman saying, *Pecunia non olet*: "Money doesn't stink." In Imperial Rome servicepeople took wastes away from public spaces and the toilets of the wealthy and piled them outside the city. Agriculture and tree-felling drained soils of nutrients and led to erosion, and the landscape became drier and more arid, with less fertile cropland. Rome's imperialism—and imperialism in general—emerged in part in response to nutrient losses, the center expanding to support its vast needs with timber, food, and other resources elsewhere. (Tellingly, as the city's resources shrank and conquests grew, Rome's agricultural deity, Mars, became the god of war.)

William Cronon chronicles a similar relationship between a city and its natural environment in *Nature's Metropolis*. He points out that the great rural areas around Chicago, America's "breadbasket," were actually organized over time to provide services for that city; the settlement of the surrounding frontier did not happen in isolation from Chicago but was inextricably bound to the city and fueled by its needs. "The central story of the nineteenth-century West is that of an expanding metropolitan economy creating ever more elaborate and intimate linkages between city and country," Cronon observes. Thus the history of a city "must also be the history of its human countryside, and of the natural world within which city and country are both located."

As they swelled and grew, the great cities placed incredible pressure on the environment around them, sucking materials and resources from farther and farther away, as the land was stripped and resources taken. For example, as the forests of Minnesota disappeared, logging moved on to British Columbia. (Such expansions affected native people; the Mandans of the upper Missouri were wiped out by smallpox, in a chain of events resulting from settlers staking homesteads.)

Over time cities all over the world built up an infrastructure for transferring nutrients from place to place. Cultures went into conflict with other cultures for resources, land, and food. In the nineteenth and early twentieth centuries, synthetic fertilizers were developed, laying the ground for the massively intensified production of industrialized agriculture. Soils now yield more crops than they naturally could, but with some severe effects: they are eroding at an unprecedented rate, and they are drained of nutrient-rich humus. Very few small farmers return local biological wastes to the soil as a primary source of nutrients any longer, and industrialized farming almost never does. Moreover, the synthetic fertilizers were often heavily contaminated with cadmium and radioactive elements from phosphate rocks, a hazard of which farmers and residents were generally unaware.

Yet certain traditional cultures have well understood the value of nutrient flows. For centuries in Egypt, the Nile River overflowed its banks each year, leaving a rich layer of silt across the valleys when waters withdrew. Beginning about 3200 B.C., farmers in Egypt structured a series of irrigation ditches that channeled the Nile's fertile waters to their fields.

They also learned to store food surpluses for periods of drought. The Egyptians maximized these nutrient flows for centuries without overtaxing them. Gradually, as British and French engineers entered the country during the nineteenth century, Egypt's agriculture shifted to Western methods. Since the completion of the Aswan High Dam in 1971, the silt that enriched Egypt for centuries now accumulates behind concrete, and people in Egypt build housing on once fertile areas originally reserved for crops. Houses and roads compete dramatically for space with agriculture. Egypt produces less than 50 percent of its own food and depends on imports from Europe and the United States.

Over thousands of years, the Chinese perfected a system that prevents pathogens from contaminating the food chain, and fertilized rice paddies with biological wastes, including sewage. Even today some rural households expect dinner guests to "return" nutrients in this way before they leave, and it is a common practice for farmers to pay households to fill boxes with their bodily wastes. But today the Chinese, too, have turned to systems based on the Western model. And, like Egypt, they are growing more dependent on imported foods.

Humans are the only species that takes from the soil vast quantities of nutrients needed for biological processes but rarely puts them back in a usable form. Our systems are no longer designed to return nutrients in this way, except on small, local levels. Harvesting methods like clear-cutting precipitate soil erosion, and chemical processes used in both agriculture and manufacture often lead to salinization and acidification, helping to deplete more than twenty times as much soil each year as nature creates. It can take approximately five hundred years for soil to build up an inch of its rich layers of microorganisms and nutrient flows, and right now we are losing five thousand times more soil than is being made.

In preindustrial culture, people did consume things. Most products would safely biodegrade once they were thrown away, buried, or burned. Metals were the exception: these were seen as highly valuable and were melted down and reused. (They were actually what we call early technical nutrients.) But as industrialization advanced, the consumption mode persisted, even though most manufactured items could no longer actually be consumed. In times of scarcity, a recognition of the value of technical materials would flare up; people who grew up during the Great Depression, for example, were careful about reusing jars, jugs, and aluminum foil, and during World War II, people saved rubber bands, aluminum foil, steel, and other materials to feed industrial needs. But as cheaper materials and new synthetics flooded the postwar market, it became less expensive for industries to make a new aluminum, plastic, or glass bottle or package at a central plant and ship it out than to build up local infrastructures for collecting, transporting, cleaning, and processing things for reuse. Similarly, in the early decades of industrialization, people might pass down, repair, or sell old service products like ovens, refrigerators, and phones to junk dealers. Today most so-called durables are tossed. (Who on Earth would repair a cheap toaster today? It is much easier to buy a new one than it is to send the parts back to the manufacturer or track down someone to repair it locally.) Throwaway products have become the norm.

There is no way, for example, that you are going to consume your car; and although it is made of valuable technical materials, you can't do anything with them once you finish with it (unless you are a junk artist). As we have mentioned, these materials are lost or degraded even in "recycling" because cars are not designed from the beginning for effective, optimal recycling as technical nutrients. Indeed, industries design products with built-in obsolescence—that is, to last until approximately the time customers typically want to replace them. Even things with a real consumable potential, such as packaging materials, are often deliberately designed not to break down under natural conditions. In fact, packaging may last far longer than the product it protected. In places where resources are hard to get, people still creatively reuse materials to make new products (such as using old tire rubber to make sandals) and even energy (burning synthetic materials for fuel). Such creativity is natural and adaptive and can be a vital part of material cycles. But as long as these uses are ignored by current industrial design and manufacturing, which typically refrain from embracing any vision of a product's further life, such reuse will often be unsafe, even lethal.

Monstrous Hybrids

Mountains of waste rising in landfills are a growing concern, but the quantity of these wastes—the space they take up—is not the major problem of cradle-to-grave designs. Of greater concern are the nutrients—valuable "food" for both industry and nature—that are contaminated, wasted, or lost. They are lost not only for lack of adequate systems of retrieval; they are lost also because many products are what we jokingly refer to as "Frankenstein products" or (with apologies to Jane Jacobs) "monstrous hybrids"—mixtures of materials both technical and biological, neither of which can be salvaged after their current lives.

A conventional leather shoe is a monstrous hybrid. At one time, shoes were tanned with vegetable chemicals, which were relatively safe, so the wastes from their manufacture posed no real problem. The shoe could biodegrade after its useful life or be safely burned. But vegetable tanning required that trees be harvested for their tannins. As a result, shoes took a long time to make, and they were expensive. In the past forty years, vegetable tanning has been replaced with chromium tanning, which is faster and cheaper. But chromium is rare and valuable for industries, and in some forms it is carcinogenic. Today shoes are often tanned in developing countries where few if any precautions are taken to protect people and ecosystems from chromium exposure; manufacturing wastes may be dumped into nearby bodies of water or incinerated, either of which distributes toxins (often disproportionately in low-income areas). Conventional rubber shoe soles, moreover, usually contain lead and plastics. As the shoe is worn, particles of it degrade into the atmosphere and soil. It cannot be safely consumed, either by you or by the environment. After use, its valuable materials, both biological and technical, are usually lost in a landfill.

A Confusion of Flows

There may be no more potent image of disagreeable waste than sewage. It is a kind of waste people are happy to get "away" from. Before modern sewage systems, people in cities would dump their wastes outside (which might mean out the window), bury them, slop them into cesspools at the bottom of a house, or dispose of them in bodies of water, sometimes upstream from drinking sources. It wasn't until the late nineteenth century that people began to make the connection between sanitation and public health, which provided the impetus for more sophisticated sewage treatment. Engineers saw pipes taking storm water to rivers and realized this would be a convenient way to remove waterborne sewage. But that didn't end the problem. From time to time the disposal of raw sewage in rivers close to home became unbearable; during the Great Stink of London in 1858, for example, the reek of raw sewage in the nearby Thames disrupted sittings of the House of Commons. Eventually, sewage treatment plants were built to treat effluents and sized to accommodate waterborne sewage combined with added storm water during major rains.

The original idea was to take relatively active biologically based sewage, principally from humans (urine and excrement, the kind of waste that has interacted with the natural world for millennia), and render it harmless. Sewage treatment was a process of microbial and bacterial digestion. The solids were removed as sludge, and the remaining liquid, which had brought the sewage to treatment in the first place, could be released essentially as water. That was the original strategy. But once the volume of sewage overwhelmed the waterways into which it flowed, harsh chemical treatments like chlorination were added to manage the process. At the same time, new products were being marketed for household use that were never designed with sewage treatment plants (or aquatic ecosystems) in mind. In addition to biological wastes, people began to pour all kinds of things down the drain: cans of paint, harsh chemicals to unclog pipes, bleach, paint thinners, nail-polish removers. And the waste itself now carried antibiotics and even estrogens from birth control pills. Add the various industrial wastes, cleaners, chemicals, and other substances that will join household wastes, and you have highly complex mixtures of chemical and biological substances that still go by the name of sewage. Antimicrobial products—like many soaps currently marketed for bathroom use—may sound desirable, but they are a problematic addition to a system that relies on microbes to be effective. Combine them with antibiotics and other antibacterial ingredients, and you may even set in motion a program to create hyperresistant superbacteria.

Recent studies have found hormones, endocrine disrupters, and other dangerous compounds in bodies of water that receive "treated" sewage effluents. These substances can contaminate natural systems and drinking-water supplies and, as we have noted, can lead to mutations of aquatic and animal life. Nor have the sewage pipes themselves been designed for biological systems; they contain materials and coatings that could degrade and contaminate effluents. As a result, even efforts to reuse sewage sludge for fertilizer have been hampered by farmers' concern over toxification of the soil.

If we are going to design systems of effluents that go back into the environment, then perhaps we ought to move back upstream and think of all the things that are designed to go into such systems as part of nutrient flows. For example, the mineral phosphate is used as a fertilizer for crops around the world. Typical fertilizer uses phosphate that is mined from rock, however, and extracting it is extremely destructive to the environment. But phosphate also occurs naturally in sewage sludge and other organic wastes. In fact, in European sewage sludge, which is often landfilled, phosphate occurs in higher concentrations than it does in some phosphate rock in China, where much of it is mined to devastating effect on local ecosystems. What if we could design a system that safely captured the phosphate already in circulation, rather than discarding it as sludge?

From Cradle-to-Grave to Cradle-to-Cradle

People involved in industry, design, environmentalism, and related fields often refer to a product's "life cycle." Of course, very few products are actually living, but in a sense we project our vitality—and our mortality—onto them. They are something like family members to us. We want them to live with us, to belong to us. In Western society, people have graves, and so do products. We enjoy the idea of ourselves as powerful, unique individuals; and we like to buy things that are brand-new, made of materials that are "virgin." Opening a new product is a kind of metaphorical defloration: "This virgin product is mine, for the very first time. When I am finished with it (special, unique person that I am), everyone is. It is history." Industries design and plan according to this mind-set.

We recognize and understand the value of feeling special, even unique. But with materials, it makes sense to celebrate the sameness and commonality that permit us to enjoy them—in special, even unique, products—more than once. What would have happened, we sometimes wonder, if the Industrial Revolution had taken place in societies that emphasize the community over the individual, and where people believed not in a cradle-to-grave life cycle but in reincarnation?

A World of Two Metabolisms

The overarching design framework we exist within has two essential elements: mass (the Earth) and energy (the sun). Nothing goes in or out of the planetary system except for heat and the occasional meteorite. Otherwise, for our practical purposes, the system is closed, and its basic elements are valuable and finite. Whatever is naturally here is all we have. Whatever humans make does not go "away."

If our systems contaminate Earth's biological mass and continue to throw away technical materials (such as metals) or render them useless, we will indeed live in a world of limits, where production and consumption are restrained, and the Earth will literally become a grave.

If humans are truly going to prosper, we will have to learn to imitate nature's highly effective cradle-to-cradle system of nutrient flow and metabolism, in which the very concept of waste does not exist. *To eliminate the concept of waste means to design things—products, packaging, and systems—from the very beginning on the understanding that waste does not exist.* It means that the valuable nutrients contained in the materials shape and determine the design: form follows evolution, not just function. We think this is a more robust prospect than the current way of making things.

As we have indicated, there are two discrete metabolisms on the planet. The first is the biological metabolism, or the biosphere—the cycles of nature. The second is the technical metabolism, or the technosphere—the cycles of industry, including the harvesting of technical materials from natural places. With the right design, all of the products and materials manufactured by industry will safely feed these two metabolisms, providing nourishment for something new.

Products can be composed either of materials that biodegrade and become food for *biological cycles*, or of technical materials that stay in closed-loop *technical cycles*, in which they continually circulate as valuable nutrients for industry. In order for these two metabolisms to remain healthy, valuable, and successful, great care must be taken to avoid contaminating one with the other. Things that go into the organic metabolism must not contain mutagens, carcinogens, persistent toxins, or other substances that accumulate in natural systems to damaging effect. (Some materials that would damage the biological metabolism, however, could be safely handled by the technical metabolism.) By the same token, biological nutrients are not designed to be fed into the technical metabolism, where they would not only be lost to the biosphere but would weaken the quality of technical materials or make their retrieval and reuse more complicated.

The Biological Metabolism

A *biological nutrient* is a material or product that is designed to return to the biological cycle—it is literally consumed by microorganisms in the soil and by other animals. Most packaging (which makes up about 50 percent of the volume of the municipal solid waste stream) can be designed as biological nutrients, what we call *products of consumption*. The idea is to compose these products of materials that can be tossed on the ground or compost heap to safely biodegrade after use—literally to be consumed. There is no need for shampoo bottles, toothpaste tubes, yogurt and ice-cream cartons, juice containers, and other packaging to last decades (or even centuries) longer than what came inside them. Why should individuals and communities be burdened with downcycling or landfilling such material? Worry-free packaging could safely decompose, or be gathered and used as fertilizer, bringing nutrients back to the soil. Shoe soles could degrade to enrich the environment. Soaps and other liquid cleaning products could be designed as biological nutrients as well; that way, when they wash down the drain, pass through a wetland, and end up in a lake or river, they support the balance of the ecosystem.

In the early 1990s the two of us were asked by DesignTex, a division of Steelcase, to conceive and create a compostable upholstery fabric, working with the Swiss textile mill Röhner. We were asked to focus on creating an aesthetically unique fabric that was also environmentally intelligent. DesignTex first proposed that we consider cotton combined with PET (polyethylene terephthalate) fibers from recycled soda bottles. What could be better for the environment, they thought, than a product that combined a "natural" material with a "recycled" one? Such hybrid material had the additional apparent advantages of being readily available, market-tested, durable, and cheap.

But when we looked carefully at the potential long-term design legacy, we discovered some disturbing facts. First, as we have mentioned, upholstery abrades during normal use, and so our design had to allow for the possibility that particles might be inhaled or swallowed. PET is covered with synthetic dyes and chemicals and contains other questionable substances—not exactly what you want to breathe or eat. Furthermore, the fabric would not be able to continue after its useful life as either a technical or a biological nutrient. The PET (from the plastic bottles) would not go back to the soil safely, and the cotton could not be circulated in industrial cycles. The combination would be yet another monstrous hybrid, adding junk to a landfill, and it might also be dangerous. This was not a product worth making.

We made clear to our client our intention to create a product that would enter either the biological or the technical metabolism, and the challenge crystallized for both of us. The team decided to design a fabric that would be safe enough to eat: it would not harm people who breathed it in, and it would not harm natural systems after its disposal. In fact, as a biological nutrient, it would nourish nature.

The textile mill that was chosen to produce the fabric was quite clean by accepted environmental standards, one of the best in Europe, yet it had an interesting dilemma. Although the mill's director, Albin Kaelin, had been diligent about reducing levels of dangerous emissions, government regulators had recently defined the mill's fabric trimmings as hazardous waste. The director had been told that he could no longer bury or burn these trimmings in hazardous-waste incinerators in Switzerland but had to export them to Spain for disposal. (Note the paradoxes here: the trimmings of a fabric are not to be buried or disposed of without expensive precaution, or must be exported "safely" to another location, but the material itself can still be sold as safe for installation in an office or home.) We hoped for a different fate for our trimmings: to provide mulch for the local garden club, with the help of sun, water, and hungry microorganisms.

The mill interviewed people living in wheelchairs and discovered that their most important needs in seating fabric were that it be strong and that it "breathe." The team decided on a mixture of safe, pesticide-free plant and animal fibers for the fabric: wool, which provides insulation in winter and summer, and ramie, which wicks moisture away. Together these fibers would make for a strong and comfortable fabric. Then we began working on the most difficult aspect of the design: the finishes, dyes, and other process chemicals. Instead of filtering out mutagens, carcinogens, endocrine disrupters, persistent toxins, and bioaccumulative substances at the end of the process, we would filter them out at the beginning. In fact, we would go beyond designing a fabric that would do no harm; we would design one that was nutritious.

Sixty chemical companies declined the invitation to join the project, uncomfortable at the idea of exposing their chemistry to the kind of scrutiny it would require. Finally one European company agreed to join. With its help, we eliminated from consideration almost eight thousand chemicals that are commonly used in the textile industry; we also thereby eliminated the need for additives and corrective processes. Not using a given dye, for example, removed the need for additional toxic chemicals and processes to ensure ultraviolet-light stabilization (that is, colorfastness). Then we looked for ingredients that had *positive* qualities. We ended up selecting only thirty-eight of them, from which we created the entire fabric line. What might seem like an expensive and laborious research process turned out to solve multiple problems and to contribute to a higher-quality product that was ultimately more economical.

The fabric went into production. The factory director later told us that when regulators came on their rounds and tested the effluent (the water coming out of the factory), they thought their instruments were broken. They could not identify any pollutants, not even elements they knew were in the water when it came into the factory. To confirm that their testing equipment was actually in working order, they checked the influent from the town's water mains. The equipment was fine; it was simply that by most parameters the water coming out of the factory was as clean as—or even cleaner than—the water going in. When a factory's effluent is cleaner than its influent, it might well prefer to use its effluent as influent. Being designed into the manufacturing process, this dividend is free and requires no enforcement to continue or to exploit. Not only did our new design process bypass the traditional responses to environmental problems (reduce, reuse, recycle), it also eliminated the need for regulation, something that any businessperson will appreciate as extremely valuable.

The process had additional positive side effects. Employees began to use, for recreation and additional work space, rooms that were previously reserved for hazardous-chemical storage. Regulatory paperwork was eliminated. Workers stopped wearing the gloves and masks that had given them a thin veil of protection against workplace toxins. The mill's products became so successful that it faced a new problem: financial success, just the kind of problem businesses want to have.

As a biological nutrient, the fabric embodied the kind of fecundity we find in nature's work. After customers finished using it, they could simply tear the fabric off the chair frame and throw it onto the soil or compost heap without feeling bad—even, perhaps, with a kind of relish. Throwing something away can be fun, let's admit it; and giving a guilt-free gift to the natural world is an incomparable pleasure.

The Technical Metabolism

A *technical nutrient* is a material or product that is designed to go back into the technical cycle, into the industrial metabolism from which it came. The average television we analyzed, for example, was made of 4,360 chemicals. Some of them are toxic, but others are valuable nutrients for industry that are wasted when the television ends up in a landfill. Isolating them from biological nutrients allows them to be *upcycled* rather than recycled—to retain their high quality in a closed-loop industrial cycle. Thus a sturdy plastic computer case, for example, will continually circulate as a sturdy plastic computer case—or as some other high-quality product, like a car part or a medical device—instead of being downcycled into soundproof barriers and flowerpots.

Henry Ford practiced an early form of upcycling when he had Model A trucks shipped in crates that became the vehicle's floorboards when it reached its destination. We are initiating a similar practice that is a modest beginning: Korean rice husks used as packing for stereo components and electronics sent to Europe, then reused there as a material for making bricks. (Rice husks contain a high percentage of silica.) The packing material is nontoxic (rice husks are safer than recycled newspapers, which contain toxic inks and particles that contaminate indoor air); its shipping is inclusive in the freight costs the electronic goods would incur anyway; and the concept of waste is eliminated.

Industrial mass can be specifically designed to retain its high quality for multiple uses. Currently, when an automobile is discarded, its component steel is recycled as an amalgam of all its steel parts, along with the various steel alloys of other products. The car is crushed, pressed, and processed so that high-ductile steel from the body and stainless steels are smelted together with various other scrap steels and materials, compromising their high quality and drastically restricting their further use. (It can't, for example, be used to make car bodies again.) The copper in its cables is melded into a general compound and lost to specific technical purposes—it can no longer be used as a copper cable. A more prosperous design would allow the car to be used the way Native Americans used a buffalo carcass, optimizing every element, from tongue to tail. Metals would be smelted only with like metals, to retain their high quality; likewise for plastics.

In order for such a scenario to be practical, however, we have to introduce a concept that goes hand in hand with the notion of a technical nutrient: the concept of a *product of service*. Instead of assuming that all products are to be bought, owned, and disposed of by "consumers," products containing valuable technical nutrients—cars, televisions, carpeting, computers, and refrigerators, for example—would be reconceived as *services* people want to enjoy. In this scenario, customers (a more apt term for the users of these products) would effectively purchase the service of such a product for a *defined user period*—say, ten thousand hours of television viewing, rather than the television itself. They would not be paying for complex materials that they won't be able to use after a product's current life. When they finish with the product, or are simply ready to upgrade to a newer version, the manufacturer replaces it, taking the old model back, breaking it down, and using its complex materials as food for new products. The customers would receive the services they need for as long as they need them and could upgrade as often as desired; manufacturers would continue to grow and develop while retaining ownership of their materials.

A number of years ago we worked on a "rent-a-solvent" concept for a chemical company. A solvent is a chemical that is used to remove grease, for example, from machine parts. Companies ordinarily buy the cheapest degreasing solvent available, even if it comes from halfway around the globe. After its use, the waste solvent is either evaporated or entered into a waste treatment flow, to be handled by a sewage treatment plant. The idea behind rent-a-solvent was to provide a degreasing service using high-quality solvents available to customers without selling the solvent itself; the provider would recapture the emissions and separate the solvent from the grease so that it would be available for continuous reuse. Under these circumstances, the company had incentive to use high-quality solvents (how else to retain customers?) and to reuse it, with the important side effect of keeping toxic materials out of waste flows. Dow Chemical has experimented with this concept in Europe, and DuPont is taking up this idea vigorously.

This scenario has tremendous implications for industry's material wealth. When customers finish with a traditional carpet, for example, they must pay to have it removed. At that point its materials are a liability, not an asset—they are a heap of petrochemicals and other potentially toxic substances that must be toted to a landfill. This linear, cradle-to-grave life cycle has several negative consequences for both people and industry. The energy, effort, and materials that were put into manufacturing the carpet are lost to the manufacturer once the customer purchases it. Millions of pounds of potential nutrients for the carpet industry alone are wasted each year, and new raw materials must continually be extracted. Customers who decide they want or need new carpeting are inconvenienced, financially burdened with a new purchase (the cost of the unrecoverable materials must be built into the price), and, if they are environmentally concerned, taxed with guilt as well about disposing of the old and purchasing the new.

Carpet companies have been among the first industries to adopt our product-of-service or "eco-leasing" concepts, but so far they have applied them to conventionally designed products. An average commercial carpet consists of nylon fibers backed with fiberglass and PVC. After the product's useful life, a manufacturer typically downcycles it—shaves off some of the nylon material for further use and discards the leftover material "soup." Alternately, the manufacturer may chop up the whole thing, remelt it, and use it to make more carpet backing. Such a carpet was not originally designed to be recycled and is being forced into another cycle for which it is not ideally suited. But carpeting designed as a true technical nutrient would be made of safe materials designed to be truly recycled as raw material for fresh carpeting, and the delivery system for its service would cost the same as or less than buying it. One of our ideas for a new design would combine a durable bottom layer with a detachable top. When a customer wants to replace the carpeting, the manufacturer simply removes the top, snaps down a fresh one in the desired color, and takes the old one back as food for further carpeting.

• • •

Under this scenario, people could indulge their hunger for new products as often as they wish, without guilt, and industry could encourage them to do so with impunity, knowing that both sides are supporting the technical metabolism in the process. Automobile manufacturers would *want* people to turn in their old cars in order to regain valuable industrial nutrients. Instead of waving industrial resources good-bye as the customer drives off in a new car, never to enter the dealership again, automobile companies could develop lasting and valuable relationships that enhance customers' quality of life for many decades and that continually enrich the industry itself with industrial "food."

Designing products as products of service means designing them to be disassembled. Industry need not design what it makes to be durable beyond a certain amount of time, any more than nature does. The durability of many current products could even be seen as a kind of intergenerational tyranny. Maybe we want our things to live forever, but what do future generations want? What about their right to the pursuit of life, liberty, and happiness, to a celebration of their own abundance of nutrients, of materials, of delight? Manufacturers would, however, have permanent responsibility for storing and, if it is possible to do so safely, reusing whatever potentially hazardous materials their products contain. What better incentive to evolve a design that does without the hazardous materials entirely?

The advantages of this system, when fully implemented, would be threefold: it would produce no useless and potentially dangerous waste; it would save manufacturers billions of dollars in valuable materials over time; and, because nutrients for new products are constantly circulated, it would diminish the extraction of raw materials (such as petrochemicals) and the manufacture of potentially disruptive materials, such as PVC, and eventually phase them out, resulting in more savings to the manufacturer and enormous benefit to the environment.

A number of products are already being designed as biological and technical nutrients. But for the foreseeable future, many products will still not fit either category, a potentially dangerous situation. In addition, certain products cannot be confined to one metabolism exclusively because of the way they are used in the world. These products demand special attention.

When Worlds Collide

If a product must, for the time being, remain a "monstrous hybrid," it may take extra ingenuity to design and market it to have positive consequences for both the biological and technical metabolisms. Consider the unintended design legacy of the average pair of running shoes, something many of us own. While you are going for your walk or run, an activity that supposedly contributes to your health and well-being, each pounding of your shoes releases into the environment tiny particles containing chemicals that may be teratogens, carcinogens, or other substances that can reduce fertility and inhibit the oxidizing properties of cells. The next rain will wash these particles into the plants and soil around the road. (If the soles of your athletic shoes contain a special bubble filled with gases for cushioning—some of which were recently discovered factors in global warming—you may also be contributing to climate change.) Running shoes can be redesigned so that their soles are biological nutrients. Then when they breaks down under pounding feet, they will nourish the organic metabolism instead of poisoning it. As long as the uppers remain technical nutrients, however, the shoes would be designed for easy disassembly in order to be safely recirculated in both cycles (with the technical materials to be retrieved by the manufacturer). Retrieving technical nutrients from the shoes of famous athletes—and advertising the fact—could give an athletic-gear company a competitive edge.

Some materials do not fit into either the organic or technical metabolism because they contain materials that are hazardous. We call them *unmarketables*, and until technological ways of detoxifying them—or doing without them—have been developed, they also require creative measures. They can be stored in "parking lots"—safe repositories that the producer of the material either maintains or pays a storage fee to use. Current unmarketables can be recalled for safe storage, until they can be detoxified and returned as valuable molecules to a safe human use. Nuclear waste is clearly an unmarketable; in a pure sense, the definition should also include materials known to have hazardous components. PVC is one such example: instead of being incinerated or landfilled, it might instead be safely "parked" until cost-effective detoxification technologies have evolved. As currently made, PET, with its antimony content, is another unmarketable: with some technological ingenuity, items that contain PET, such as soda bottles, might even be upcycled to remove the antimony residues and to create a clean polymer ready for continuous, safe reuse.

Companies might undertake a *waste phaseout*, in which unmarketables—problematic wastes and nutrients—are removed from the current waste stream. Certain polyesters now on the market could be gathered and their problematic antimony removed. This would be preferable to leaving them in textiles, where they will eventually be disposed of or incinerated, perhaps therefore to enter natural systems and nutrient flows. The materials in certain monstrous hybrids could be similarly gathered and separated. Cotton could be composted out of polyester-cotton textile blends, and the polyester then returned to technical cycles. Shoe companies might recover chromium from shoes. Other industries might retrieve parts of television sets and other service products from landfills. Making a successful transition requires leadership in these areas as well as creative owning up.

Should manufacturers of existing products feel guilty about their complicity in this heretofore destructive agenda? Yes. No. It doesn't matter. Insanity has been defined as doing the same thing over and over and expecting a different outcome. Negligence is described as doing the same thing over and over even though you know it is dangerous, stupid, or wrong. Now that we know, it's time for a change. Negligence starts tomorrow.

Chapter Five

Respect Diversity

Imagine the primordial beginning of life on this planet. There is rock and water—matter. The orb of the sun sends out heat and light—energy. Eventually, over thousands of millennia, through chemical and physical processes scientists still don't fully understand, single-celled bacteria emerge. With the evolution of photosynthesizing blue-green algae, a monumental change takes place. Chemistry and physics combine with the sun's physical energy, and the Earth's chemical mass turns into the blue-green planet we know.

Now biological systems evolve to feed on energy from the sun, and all heaven breaks loose. The planet's surface explodes with life forms, a web of diverse organisms, plants, and animals, some of which, billions of years later, will inspire powerful religions, discover cures for fatal diseases, and write great poems. Even if some natural disaster occurs—if, say, an ice age freezes large parts of the earth's surface—this pattern is not destroyed. As the ice retreats, life creeps back. In the tropics, a volcano erupts and smothers the surrounding land in ash. But a coconut shell floats across waters and ends up as debris on a beach, or a spore or spiderling moves through air, lands on a crumbling rock, and begins to reweave nature's web. It's a mysterious process, but a miraculously stubborn one. When faced with blankness, nature rises to fill in the space.

This is nature's design framework: a flowering of diversity, a flowering of abundance. It is Earth's response to its one source of incoming energy: the sun.

The current design response of humans to this framework might be called "attack of the one-size-fits-all." Layers of concrete and asphalt obliterate forests, deserts, coastal marshes, jungles—everything in their path. Buildings that present a bland, uniform front rise in communities where structures were for decades, even centuries, beautiful and culturally distinct. Spaces once lush with foliage and wildlife shrink to marginal places where only the hardiest species—crows, roaches, mice, pigeons, squirrels—survive. Landscapes are flattened into lawns of a single species of grass, artificially encouraged to grow but constantly cut back, with controlled hedges and a few severely pruned trees. The monotony spreads and spreads, overwhelming the details of place in its path. What it seems to seek is simply more of itself.

We see this as *de-evolution*—simplification on a mass scale—and it is not limited to ecology. For centuries, our species has built up a variety of cultures across the globe, ways of eating, speaking, dressing, worshiping, expressing, creating. A tide of sameness is spreading from sea to sea, sweeping away these cultural details too.

Against this tide of sameness we advance the principle "respect diversity." By this we mean to include not only biodiversity but also diversity of place and of culture, of desire and need, the uniquely human element. How can a factory built in a desert climate be delightfully different from one constructed in the tropics? What does it mean to be Balinese, to be Mexican, and to express it? How can we enrich local species, and invite them into our "cultivated" landscapes instead of destroying or chasing them away? How can we gain profit and pleasure from a diversity of natural energy flows? How can we engage with an abundance of diverse materials, options, and responses, of creative and elegant solutions?

The Fittest Survive, the Fitting-est Thrive

Popular wisdom holds that the fittest survive, the strongest, leanest, largest, perhaps meanest—whatever beats the competition. But in healthy, thriving natural systems it is actually the *fitting-est* who thrive. Fitting-est implies an energetic and material engagement with place, and an interdependent relationship to it.

Think again of the ants. We may have an archetypal notion of "ant," but in fact there are more than eight thousand different kinds of ants that inhabit the planet. Over millions of years, each has evolved to fit its particular locale, developing features and behaviors that enable it to carve out a habitat and to cull the energy and nourishment it needs. In the rain forest, hundreds of different species of ants may coexist in the crown of a single large tree. There is the leaf-cutter ant, with mandibles designed to cut and carry foliage; the fire ant, a scavenger with advanced methods of group transport to tote prey of various sizes to its nest; the weaver ant, with its advanced pheromone communication system used to call allies and workers to war;

the trap-jaw ant, whose ferocious snapping jaw is legendary. Around the world there are ants that hunt alone, ants that hunt in groups, and ants that raise broods of aphid "cattle," which they milk for sweet liquid. In a startling use of solar power, hundreds of one colony's workers may cluster on the forest floor to soak up sunlight before carrying its warmth in their very bodies back down to the nest.

Being fitting, ants do not inevitably work to destroy competing species. Rather, they compete productively from their niches, the term scientists use to describe species' various zones of habitation and resource use within an ecosystem. In his book *Diversity and the Rain Forest*, John Terborgh, a scientist who has studied the complex ecosystems of the rain forest, explains how ten species of ant wren manage to cohabit a single area of the forest while preying on the same kinds of insects: one species inhabits an area close to the ground, several more live in the middle tiers of the trees, and another occupies the high canopy. In each of these areas, species forage differently—one middle-tier wren gleans the leaves for insects, another the twigs and branches, and so forth, leaving food in the other niches.

The vitality of ecosystems depends on relationships: what goes on between species, their uses and exchanges of materials and energy in a given place. A tapestry is the metaphor often invoked to describe diversity, a richly textured web of individual species woven together with interlocking tasks. In such a setting, diversity means strength, and monoculture means weakness. Remove the threads, one by one, and an ecosystem becomes less stable, less able to withstand natural catastrophe and disease, less able to stay healthy and to evolve over time. The more diversity there is, the more productive functions—for the ecosystem, for the planet—are performed.

Each inhabitant of an ecosystem is therefore interdependent to some extent with the others. Every creature is involved in maintaining the entire system; all of them work in creative and ultimately effective ways for the success of the whole. The leaf-cutter ants, for example, recycle nutrients, taking them to deeper soil layers so that plants, worms, and microorganisms can process them, all in the course of gathering and storing food for themselves. Ants everywhere loosen and aerate the soil around plant roots, helping to make it permeable to water. Trees transpire and purify water, make oxygen, and cool the planet's surface. Each species' industry has not only individual and local implications but global ones as well. (In fact, some people, such as those who subscribe to the Gaia principle, go so far as to perceive the world as a single giant organism.)

If nature is our model, what does it mean for human industries to be involved in maintaining and enriching this vibrant tapestry? First, it means that in the course of our individual activities, we work toward a rich connection with place, and not simply with surrounding ecosystems; biodiversity is only one aspect of diversity. Industries that respect diversity engage with local material and energy flows, and with local social, cultural, and economic forces, instead of viewing themselves as autonomous entities, unconnected to the culture or landscape around them.

All Sustainability Is Local

We begin to make human systems and industries fitting when we recognize that all sustainability (just like all politics) is local. We connect them to local material and energy flows, and to local customs, needs, and tastes, from the level of the molecule to the level of the region itself. We consider how the chemicals we use affect local water and soil—rather than contaminate, how might they nourish?—what the product is made from, the surroundings in which it is made, how our processes interact with what is happening upstream and downstream, how we can create meaningful occupations, enhance the region's economic and physical health, accrue biological and technical wealth for the future. If we import a material from a distant place, we honor what happened there as a local event. As we wrote in *The Hannover Principles*, "Recognize interdependence. The elements of human design are entwined with and depend upon the natural world, with broad and diverse implications at every scale. Expand design considerations and recognize distant effects."

When Bill traveled to Jordan with his professor in 1973 to work on a long-term plan for the future of the East Bank of the Jordan River Valley, the team's design assignment was to identify strategies for building towns of the future in which the Bedouin could settle, now that political borders had put a stop to their traditional nomadic migrations. A competing team proposed Soviet-style prefabricated housing blocks of a sort that became ubiquitous in the former Eastern Bloc and USSR, "anywhere" buildings that can be found from Siberia to the

Caspian Desert. The buildings themselves would be trucked down rough roads from an industrial center in the highlands near Amman and assembled in the valley.

Bill and his colleagues created a proposal to adapt and encourage adobe structures. Local people could build these with materials at hand—clay and straw, horse, camel, or goat hair, and (not least) abundant sun. The materials were ancient, well understood, and uniquely suited to the hot, dry climate. The structures themselves were designed to optimize temperature flux over the course of the day and year: at night their mass absorbed and stored the coolness of the air, which would keep the interior temperature down during the hot desert days. The team tracked down elder craftspeople in the region who could show them how to build the structures (especially the domes) and then train the Bedouin youths (who had grown up with tents) to build with and repair adobe in the future.

The question that helped to guide the team's work at every level was: What is the right thing for this place? Not prefabricated elements, or mastery of the landscape with a universal modern style, they concluded. They hoped their plan would enhance that particular community in several ways: the homes were built from local materials that were biologically and technically reusable. Employing these materials and the services of nearby craftsmen would generate local economic activity and support as many residents as possible. It would involve local people in building the community and keep them connected to the region's cultural heritage, which the structures' aesthetic distinctiveness itself helped to perpetuate. Enlisting local craftsmen to train young people in the use of local materials and techniques would encourage an intergenerational connection.

Using Local Materials

The idea of local sustainability is not limited to materials, but it begins with them. Using local materials opens the doors to profitable local enterprise. It also avoids the problem of bioinvasion, when transfer of materials from one region to another inadvertently introduces invasive nonnative species to fragile ecosystems. Chestnut blight, responsible for wiping out chestnut trees in the United States, entered this country on a piece of lumber from China. Chestnuts were a dominant tree of the eastern forests. The other native species evolved together with them, and now they are gone.

We consider not only physical materials but physical processes and their effect on the surrounding environment. Instead of destroying a landscape with conventional hack-and-mow practices, we imagine how to invite more local species in (as we did with the Herman Miller factory). By seeing sustainability as both a local and a global event, we can understand that just as it is not viable to poison local water and air with waste, it is equally unacceptable to send it downstream, or to ship it overseas to other, less regulated shores.

Perhaps the ultimate example of effective use of local materials lies in processing what we know as human waste—a fundamental application also of the principle "waste equals food." We have been working on the creation of sewage treatment plants based on bioremediation (the breaking down and purifying of wastes by nature), to replace the conventional harsh chemical treatment of sewage. Biologist John Todd calls these systems "living machines," because they use living organisms—plants, algae, fish, shrimp, microbes, and so on—instead of toxins like chlorine to purify water. These living machines are often associated with artificial environments created in greenhouses, but they have taken all kinds of forms. Some of the systems we are currently integrating into our projects are designed to work outside and year-round, in all kinds of climates. Others are constructed wetlands, or even reed beds floating on a toxic lagoon, outfitted with little windmills to move the sludge through.

For developing countries, this approach to sewage treatment represents a huge opportunity to maximize nutrient flows and implement a nutritious agenda right away. As the tropics rapidly develop, populations are expanding, and the pressure to clean up effluents (and the bodies of water in which they are routinely disposed) increases. Instead of adopting a one-size-fits-all design solution that is highly ineffective in the long run, we are encouraging these diverse cultures to develop new sewage treatment systems that make waste equal food. In 1992 a model waste treatment system developed by Michael and his colleagues was opened at Silva Jardin, in the province of Rio, Brazil. It was locally fabricated using clay pipes that carried wastewater from village residents to a large settling tank, then into an intricately connected series of small ponds full of an astonishing diversity of plants, microbes, snails, fish, and shrimp. The system was designed to recover nutrients along the way, producing clean, safe drinking water as a by-product. Farmers competed for access to this purified water and to the sludge's valuable nitrogen, phosphorus, and trace materials as nutrients for farming. Instead of being a liability, the sewage was from the outset perceived and treated as an asset of great value.

A community we are working with in Indiana simply stores its septage (the solids from sewage) in underground tanks during the chilly winters. In the summer, when the sun shines long and bright, the septage is moved to a large outdoor garden and constructed wetland, where plants, microbes, fungi, snails, and other organisms purify and use its nutrients with the power of the sun. This system is locally relevant in several ways. It works with the seasons, optimizing solar power when it is available, instead of forcing treatment during the winter when solar heat is scarce. It uses native nutrients and plants for a process that returns quality drinking water to the aquifer and sustains a lovely garden. The community ends up with millions of sewage treatment "plants"—a living example of biodiversity.

A further point: in this case, there was only one logical site for sewage treatment, on the edge of the community next to a major highway—which happens to be upstream. Because they have kept the effects of their sewage local, residents think twice about pouring a dangerous substance down the sink, or about mixing technical and biological materials. It renders palpable to them that their effluents do matter, not in some abstract way, but to real people and their families. But even if we had been able to situate the sewage site "away," we would have done well to act as if it were right where it is. In planetary terms, we're all downstream.

Connecting to Natural Energy Flows

In the 1830s Ralph Waldo Emerson traveled to Europe on a sailboat and returned on a steamship. If we look at this moment symbolically, we could say he went over on a recyclable vessel that was solar-powered, operated by craftsmen practicing ancient arts in the open air. He returned in what would become a steel rust bucket spewing oil on the water and smoke in the sky, operated by men shoveling fossil fuels into the mouths of boilers in the dark. In his journals on the way back in the steamship, Emerson noted the lack of what he wistfully described as the connection to the "Aeolian kinetic"—the force of the wind. He wondered at the implications of these changing connections between humans and nature.

Some of those implications might well have dismayed him. With new technologies and brute force energy supplies (such as fossil fuels), the Industrial Revolution gave humans unprecedented power over nature. No longer were people so dependent on natural forces, or so helpless against the vicissitudes of land and sea. They could override nature to accomplish their goals as never before. But in the process, a massive disconnection has taken place. Modern homes, buildings, and factories, even whole cities, are so closed off from natural energy flows that they are virtual steamships. It was Le Corbusier who said the house was a machine for living in, and he glorified steamships, along with airplanes, cars, and grain elevators. In point of fact, the buildings he designed had cross-ventilation and other people-friendly elements, but as his message was taken up by the modern movement, it evolved into a machinelike sameness of design. Glass, the heroic material that could connect indoors and outdoors, was used as a way of cutting us off from nature. While the sun shone, people toiled under fluorescent lights, literally working in the dark. Our structures might be machines for living in, but there was no longer much about them that was alive. (A 1998 *Wall Street Journal* article about our buildings' novel feature of having windows that open—that being a hot new commodity—reflected a true low point in the annals of contemporary commercial architecture.)

What a far cry from the saltbox houses of colonial New England, constructed with a high south side where the house's precious windows were mostly clustered, to maximize exposure to the winter sun. (In summer, the leaves of a large maple to the southwest provided shelter from the sun.) A central fireplace and chimney mass provided a warm hearth at the heart of the home, and the low north roof huddled the heated mass away from cold behind a windrow of evergreen trees planted and maintained expressly for the purpose. The structure and the surrounding landscape worked together as a total design.

It is easy to forget, in the gas-powered glare of a postindustrial age, that not only local materials and customs but energy flows have provenance. In less industrialized parts of the world, however, creative approaches to capturing local energy flows are still very much alive. The aboriginal people on the coast of Australia have a simple, elegant strategy for harnessing sunlight: two forked sticks with a single pole across the top make a beam against which bark is laid and overlapped like roof tiles on the south side during cooler months, so the inhabitants can sit in the warm north sun. In summer, they move the bark to the north side to block the sun and sit on the other side, in the shade. Their entire "building" consists of a few sticks and bark ingeniously adapted to local circumstance.

Wind towers have been used for thousands of years in hot climates to capture airflows and draw them through dwellings. In Pakistan, chimneys topped with "wind scoops" literally scoop wind and channel it down the chimney, where there might be a small pool of water for cooling the wind as it moves downward and into the house. Iranian wind towers consist of a ventilated structure that constantly drips water; air comes in, flows down the chimney with its dripping sides, and enters the house, cooled. At Fatepur Sikri in India, porous sandstone screens, sometimes intricately carved, were saturated with water to cool air passing through. In the Loess Plains of China, people dig their homes in the ground to secure shelter from wind and sun.

But with modern industrialization and its products, such as large sheets of window glass, and the widespread adoption of fossil fuels for cheap and easy heating and cooling, such local ingenuity has faded from industrialized areas, and even in rural regions it is in decline. Oddly enough, professional architects seem to get by without understanding the basic principles that inspired ancient building and architecture orientations. When Bill gives talks to architects, he asks who knows how to find true south—not magnetic or "map" south but true solar south—and gets few or no hands (and, stranger still, no requests to learn how).

Connecting to natural flows allows us to rethink everything under the sun: the very concept of power plants, of energy, habitation, and transportation. It means merging ancient and new technologies for the most intelligent designs we have yet seen. What it doesn't mean, however, is to become "independent." The popular image of going solar is linked to the concept of "going off the grid"—becoming cut off from the current energy infrastructure. This is not at all what we are implying. First of all, a renewed connection to natural flows will of necessity be gradual, and making use of existing systems is a sensible transitional strategy. Hybrid systems can be designed to draw upon local renewable energy flows in addition to artificial sources while more optimized solutions are being developed and implemented. In some cases, solar power—and also wind and water power—can be channeled into the current system of energy supplies, greatly diminishing the load of artificial energy needed and even saving money. Is this eco-efficiency? By all means. But it is eco-efficiency as a tool in service to a larger vision, not as a goal in itself.

In the long run, connecting to natural energy flows is a matter of reestablishing our fundamental connection to the source of all good growth on the planet: the sun, that tremendous nuclear power plant 93 million miles away (exactly where we want it). Even at such distances, the sun's heat can be devastating, and it commands a healthy respect for the delicate orchestration of circumstances that makes natural energy flows possible. Humans thrive on the earth under such intense emanations of heat and light only because billions of years of evolutionary processes have created the atmosphere and surface that support our existence—the soil, plant life, and cloud cover that cool the planet down and distribute water around it, keeping the atmosphere within a temperate range that we can live in. So reestablishing our connection to the sun by definition includes maintaining interdependence with all the other ecological circumstances that make natural energy flows possible in the first place.

Here are some thoughts on—and examples of—ways of optimizing energy production and use, in which diversity plays a key role.

A Transition to Diverse and Renewing Energy Flows

Earlier, we considered how diversity makes an ecosystem more resilient and able to respond successfully to change. During times of unexpected disruption—like the summer of 2001, when unusually high energy demand in California led to rolling blackouts, skyrocketing prices, even accusations of profiteering—a more complex system can adapt and survive. The same is true of an economic system: a distributed industry makes for many small players, and a more stable, resilient system for both providers and customers. And from an eco-effective perspective, the greatest innovations in energy supply are being made by small-scale plants at the local level. For example, in our work with one utility in Indiana, it appears that producing power at the scale of one small plant for every three city blocks is dramatically more effective than more centralized production. The shorter distances reduce the power lost in high-voltage transmission to insignificant levels.

Nuclear power plants and other large-scale energy providers throw off tremendous heat energy that goes unused and often disrupts the surrounding ecosystem, as when it is cooled by way of a neighboring river. With smaller utilities, it becomes possible to harness waste heat to feed local needs. For example, the hot water generated by a small fuel cell or microturbine installed in a restaurant, say, or even a residence, can be put to immediate use, a terrific convenience (and savings) to businesses and homeowners.

Rather than install more large-scale power-generating equipment to meet peak energy loads, utility companies can integrate solar collectors as products of service with systems currently in use. Residents and businesses could be asked for permission to lease their south-facing or flat roofs for this purpose, or to access solar collectors already in place. (These roofs need not look like castoffs from the space program, by the way. The ubiquitous flat commercial roof is easy to solarize, and the least expensive solar arrays are simply laid down like tiles. In many parts of California they are cost-effective now.) During peak use times, this diversely supplied system is much more in tune with its own peaks; the highest demand on the power system is created by the desire for air-conditioning, when the sun is high—exactly when solar collectors are working best. It can meet periods of intense demand much more flexibly than centralized energy monocultures of coal, gas, and nuclear power.

Another approach to the dramatic (and expensive) fluctuations in energy demand: "intelligent" appliances that receive information about the current price of power along with the power itself, and choose from alternate power sources accordingly, like a broker instructed to buy or sell according to the rise and fall of a given stock price. Why should you be paying prime-time rates to have your refrigerator chill your milk at two o'clock on a summer afternoon, when air-conditioning use has the city on the verge of rolling blackouts? Instead, your appliance could decide—according to criteria you determine—when to buy power and when to turn to a block of eutectic salts or ice that it conveniently froze the night before, ready to keep your refrigerator cool until demand and price come down. It's back to the future: voilà, you have an icebox. But you're availing yourself of the cheapest, most readily available power for a simple process, and you're not competing with the needs of a hospital emergency room to do so.

A similar focus on diversity and immediately available resources resulted in a breakthrough in energy use in a large automotive manufacturing facility. The engineers were having a difficult time finding an affordable way to make workers comfortable. All the little things that could save money weren't adding up to much. They were working with a typical approach to heating and cooling, in which thermostats placed near burners and air-conditioning units up near the roof sensed the need to cool or heat the building. In winter, hot air rose toward the roof, drawing in cold air from outside, and had to be heated again by burners and pumped down to displace the cold air it drew in. All this motion of air created an unwelcome breeze that required even more heating to counteract.

An engineer named Tom Kiser, of Professional Supply Incorporated, proposed a radical new strategy. Rather than drilling columns of cooled or heated air (as the seasons required) down toward employees at high speed from "efficiently designed" fans and ducts at the top of the building, he suggested approaching the building itself as a giant duct. When the building was pressurized with the help of four "bigfoots"—simple large units—any holes in the structure, windows and doors, for example, could be made to pass air like pinholes in an inner tube, leaking air out rather than letting it in. This had some significant benefits. In warm weather, you could simply drop a blanket of temperate air in the building, and it would sink to the factory floor without the need for multiple air-conditioning units or high-speed fans, which would have been dramatically more expensive to operate, no matter how efficiently they were made to function. During the winter, a blanket of cool air acted as a lid, keeping the warm air generated by the factory equipment down on the floor, where people needed the heat. (Without the breeze created by excessive air motion, anything about 68 degrees Fahrenheit felt plenty warm enough.) In other words, Kiser's genius was to heat with cool air. Thermostats were placed near employees, not in the equipment up near the roof, in keeping with the idea of heating and cooling people as needed, not the building itself.

Other benefits accrued. For example, in a conventional system, the opening and closing of truck docks constantly leaks in uncomfortably hot or cold air. A pressurized system keeps undesired air at bay rather than having to cool or heat it to restore the status quo. And excess heat generated by air compressors (which lose 80 percent of the energy they use as "waste" heat), welders, and other equipment could be easily captured and consolidated for use in the bigfoots. It turns what is generally a waste and a thermal liability into a working asset.

If you combined such a system with a grass roof to insulate the structure and protect it from heat gain in the summer, wind loss in the winter, and the wear and tear of daylight, you'd be treating the building as an aerodynamic event, designing it like a machine—but this time, instead of a machine for living in, a machine that's alive.

Reap the Wind

Wind power offers similar possibilities for hybrid systems that make more effective use of local resources. In places like Chicago, the "windy city" (where we are working with Mayor Richard Daley to help create "the greenest city in the United States"), and the Buffalo Ridge, which runs along the border of Minnesota and South Dakota and is sometimes referred to as the Saudi Arabia of wind, it's not difficult to imagine what local source of potential energy is most abundant. We are already seeing multi-megawatt wind farms on the Buffalo Ridge, and the state of Minnesota has offered incentive programs for wind-farm development. The Pacific Northwest, too, now sees itself as a wind-power powerhouse, and new wind farms are springing up in Pennsylvania, Florida, and Texas. Europe has had aggressive wind-energy programs for years.

From an eco-effective perspective, however, the design of conventional wind-power plants is not always optimal. The new wind farms are huge—as many as a hundred windmills (wind turbines, actually) grouped together, each of them a Goliath capable of producing one megawatt of electricity with a blade span the length of a football field. Developers like the centralized infrastructure, but the high-powered transmission lines they require means new giant towers marching over a once bucolic landscape, in addition to the windmills themselves. Also, modern windmills are not designed as technical nutrients with ecologically intelligent materials.

Think back to those famous Dutch landscape paintings. The windmills were always located among the farms, a short distance from the fields, for convenient water pumping and milling. They were distributed across the land at a scale appropriate to it, and they were made from safe local materials—and looked beautiful to boot. Now imagine one of the new windmills distributed on every few family farms in the Great Plains. As with solar collectors, utilities could lease land from the farmers for this purpose, distributing the windmills and the power they generate in a way that optimizes existing power lines and requires few new ones. The farmers get much-needed supplemental income, and the utility gets to reap the power, which it adds to the grid. One of our projects for automotive energy conceives of wind power reaped in just this way; we call it "Ride the Wind."

Those who have difficulty imagining this becoming a major source of energy might consider what the tremendous industrial capacity that allows the United States to produce millions of automobiles per year might do if a fraction of it were applied in this direction. And with the new windmills already cost-effective and directly competitive with fossil-fuel-derived and nuclear energy in appropriate landscapes, there's no reason why it shouldn't be. Combined with intelligent applications of direct solar absorption and cost-effective conservation, the implications for national prosperity and security (thanks to sovereign sources of energy) are staggering. Just imagine the robust benefits of having a new wind-turbine industry that produces home-grown hydrogen for our pipelines and vehicles instead of being forced to rely on politically and physically fragile oil shipped in supertankers from halfway around the world.

Transitional strategies for energy use give us the opportunity to develop technology that is truly eco-effective—not less depleting but replenishing. Ultimately, we want to be designing processes and products that not only return the biological and technical nutrients they use, but pay back with interest the energy they consume.

Working with a team assembled by Professor David Orr of Oberlin College, we conceived the idea for a building and its site modeled on the way a tree works. We imagined ways that it could purify the air, create shade and habitat, enrich soil, and change with the seasons, eventually accruing more energy than it needs to operate. Features include solar panels on the roof; a grove of trees on the building's north side for wind protection and diversity; an interior designed to change and adapt to people's aesthetic and functional preferences with raised floors and leased carpeting; a pond that stores water for irrigation; a living machine inside and beside the building that uses a pond full of specially selected organisms and plants to clean the effluent; classrooms and large public rooms that face west and south to take advantage of solar gain; special windowpanes that control the amount of UV light entering the building; a restored forest on the east side of the building; and an approach to landscaping and grounds maintenance that obviates the need for pesticides or irrigation. These features are in the process of being optimized—in its first summer, the building began to generate more energy capital than it used—a small but hopeful start.

Imagine a building like a tree, a city like a forest.

A Diversity of Needs and Desires

Respecting diversity in design means considering not only how a product is made but how it is to be used, and by whom. In a cradle-to-cradle conception, it may have many uses, and many users, over time and space. An office building or store, for example, might be designed so that it can be adapted to different uses over many generations of use, instead of built for one specific purpose and later torn down or awkwardly refitted. The SoHo and TriBeCa neighborhoods in lower Manhattan continue to thrive because their buildings were designed with several enduring advantages that today would not be considered efficient: they have high ceilings and large, high windows that let in daylight, thick walls that balance daytime heat with nighttime coolness. Because of their attractive and useful design, these buildings have gone through many cycles of use, as warehouses, showrooms, and workshops, then storage and distribution centers, then artists' lofts, and, more recently, offices, galleries, and apartments. Their appeal and usefulness is enduringly apparent. Following this lead, we've designed some corporate buildings to be convertible to housing in the future.

Like the French jam pots that could be used as drinking

glasses once the jam was gone, packaging and products can be designed with their future upcycling in mind. Exterior packaging, with its premium on large, flat, stiff surfaces, is a natural precursor to a further life as building materials, as Henry Ford knew. A crate that is to be used to ship a product from Savannah could be made of waterproof insulation that recipients in Soweto would use in constructing houses. Again, cultural distinctions are part of the picture. African villagers who used to drink out of gourds or clay cups and have no recycling structure for "trash" might need a drink package that can be thrown onto the ground to decompose and provide food for nature. In India, where materials and energy are very expensive, people might welcome packaging that is safe to burn. In industrial areas, a better solution might be polymers designed as "food" for more bottles, with an appropriately designed upcycling infrastructure.

In China, Styrofoam packaging presents such a disposal problem that people often refer to it as "white pollution." It is thrown from the windows of trains and barges and litters the landscape everywhere. Imagine designing such packaging to safely biodegrade after use. It could be made from the empty rice stalks that are left in the fields after harvest, which are now usually burned. They are readily available and cheap. The packaging could be enriched with a small amount of nitrogen (potentially retrieved from automotive systems). Instead of feeling guilty and burdened when they are finished eating, people could enjoy throwing their safe, healthy nutripackage out the train window onto the ground, where it would quickly decompose and provide nitrogen to the soil. It could even contain seeds of indigenous plants that would take root as the packaging decomposes. Or people could wait to dispose of the packaging at the next train stop, where local farmers and gardeners would have set up stations to collect it for use in fertilizing crops. We could even plant signs that say "Please Litter."

Form Follows Evolution

Instead of promoting a one-size-fits-all aesthetic, industries can design in the potential for "mass" customization, allowing packaging and products to be adapted to local tastes and traditions without compromising the integrity of the product. Luxury industries like fashion and cosmetics have been the trailblazers in allowing for customization to individual taste and local custom. Others can follow their lead, accommodating the need for individual and cultural expression in their designs. For example, the automobile industry might honor the Filipino practice of decorating vehicles, providing customers with the opportunity to attach fringe and to paint creative, outrageous designs in eco-friendly paints instead of constraining them to a "universal" look (or having them lose eco-effective benefits when they assert the cultural predilection for adornment). Eco-effective design demands a coherent set of principles based on nature's laws and the opportunity for constant diversity of expression. It has been famously said that form follows function, but the possibilities are greater when form follows evolution.

What goes for aesthetics goes for needs, which vary with ecological, economic, and cultural circumstances—not to mention individual preferences. As we have pointed out, soap as it is currently manufactured is designed to work the same way in every imaginable location and ecosystem. Faced with the questionable effects of such a design, eco-efficiency advocates might tell a manufacturer to "be less bad" by shipping concentrates instead of liquid soap, or by reducing or recycling packaging. But why try to optimize the wrong system? Why this packaging in the first place? Why these ingredients? Why a liquid? Why one-size-fits-all?

Why not make soap the way the ants would? Soap manufacturers could retain centralized intelligence (the concept of "soap"), but develop local packaging, shipping, and even molecular effects. For example, shipping water (in the form of liquid detergent) increases transportation expense and is unnecessary, since there is water in the washing machine, laundry, tub, river, or lake where the washing is done. Maybe soap could be delivered in pellet or powder form and sold in bulk at the grocery store. Water needs differ in different places: different kinds of pellets and powder might be used for places with hard water or soft water, still others for places where people pound clothing on rocks, feeding soap directly into the water supply. A major soap manufacturer was beginning to think this way when it paid attention to the fact that women in India were using its soaps (which had been designed for washing machines) to wash clothing by hand, sprinkling the harsh soap onto the clothing with their fingers and then pounding the clothes on rocks at the side of the river. And the women could afford to buy only a small amount of soap at a time. Faced with competition from a more versatile product, the soap company developed a gentler product and began to produce it in small, inexpensive packets that the women could open on the spot. Such thinking can go much further. For instance, manufacturers could reconceive soap as a product of service, and design washing machines to recover detergent and use it again and again. A washing machine could be leased preloaded with two thousand loads' worth of internally recycling detergent—not nearly as big a design challenge as it sounds, since only 5 percent of a standard measure of detergent is actually consumed in a typical laundry cycle.

Biologist Tom Lovejoy tells a story about a meeting between E. O. Wilson, the great evolutionary biologist who has written extensively on biodiversity (and on ants), and George H. W. Bush's chief of staff, John Sununu, around the time of the Earth Summit in 1992. Wilson was there to encourage the president to support the Biodiversity Convention being put forward by the majority of the world's countries as a signal of their dire concern over this issue. When Wilson had finished describing the great value of biodiversity, Sununu responded, "I see. You want an endangered species act for the whole world . . . and the devil is in the details." To which Wilson responded, "No, sir. God is in the details."

When diversity is nature's design framework, human design solutions that do not respect it degrade the ecological and cultural fabric of our lives, and greatly diminish enjoyment and delight. Charles de Gaulle is reported to have said that it is difficult to manage a country that produces four hundred kinds of cheese. But what if, for the sake of market growth, all the cheese makers of France began to concentrate on producing individually wrapped squares of orange "cheese food" that all tasted exactly the same?

According to visual preference surveys, most people see culturally distinctive communities as desirable environments in which to live. When they are shown fast-food restaurants or generic-looking buildings, they score the images very low. They prefer quaint New England streets to modern suburbs, even though they may live in developments that destroyed the Main Streets in their very own hometowns. When given the opportunity, people choose something other than that which they are typically offered in most one-size-fits-all designs: the strip, the subdivision, the mall. People want diversity because it brings them more pleasure and delight. They want a world of four hundred cheeses.

Diversity enriches the quality of life in another way: the furious clash of cultural diversity can broaden perspective and inspire creative change. Think how Martin Luther King, Jr., adapted Mahatma Gandhi's teachings on peaceful transformation to the concept of civil disobedience.

A Tapestry of Information

Traditionally, companies have relied on feedback for signals that influence change, looking backward to assess previous failures and successes, or they have looked around them to discover what the competition is up to. Respecting diversity means widening the scope of input too, to embrace a broader range of ecological and social contexts and a longer temporal framework as well. We can consult "feedforward," asking ourselves not only what has worked in the past and present, but what will work in the future. What kind of world do we intend, and how might we design things in keeping with that vision? What will a sustaining global commerce look like ten—or even a hundred—years from now? How can our products and systems help to create and sustain it, so that future generations are enriched by what we make, not tyrannized by hazards and waste? What can we do now to begin the process of industrial re-evolution?

If that laundry detergent manufacturer continued to think in this direction, it would move beyond the question of creating a detergent that is convenient to use and gentler on human hands to ask, Is it gentle on the Ganges? Will it foster diverse aquatic life? Now that we know what kind of soap the customers want, what kind of soap does the river want? Now that it is packaged for individual applications, how can the packaging be designed as a product of consumption that will readily biodegrade on the riverbank, contributing nutrients to the soil, or be burned safely as fuel, or both? What about fabrics that don't need soap to get clean, that are designed to enjoy a "lotus effect"? (Nothing sticks to a lotus leaf.) One by one, the elements of a product might be redefined positively against an ever widening backdrop, until the product itself evolves and is transformed, and its every aspect is designed to nourish a diverse world.

Working with a major European soap manufacturer on a

shower gel, we set ourselves the design challenge of responding to the question, What kind of soap does the river want? (The river in question was the Rhine.) At the same time, we aimed to fulfill customers' desire for a healthy, pleasurable shower gel. In the initial approach, Michael told the manufacturer that he wanted to define the product in the way that medicine was defined, proactively choosing the best ingredients. Given the nature of the product, the client company was more receptive to this approach than, say, a chemical company manufacturing house paints might be. Michael and our colleagues identified twenty-two chemicals in a typical shower gel, a number of which were added to counteract the harsh effects of other cheap chemical ingredients. (For instance, moisturizing agents were added to offset the drying effects of a particular chemical.) Then he and the team set about selecting a far smaller list of ingredients that would have only the effects they sought, designing out the intricate checks and balances of conventional formulas and resulting in a product that would be healthy for both the skin and the ecosystem of the river where it would end up.

Once the list of proposed ingredients was compiled—a total of nine—the company initially refused to go forward with the product, because the new chemicals were more expensive than the ones it had been using. But when the company considered the entire process, not only the cost of the ingredients, it came to light that the new soap was approximately 15 percent cheaper to make, thanks to simpler preparation and storage requirements. The gel went on sale in 1998 and it is still on the market—but now in a pure polypropylene packaging after Michael and the researchers found that antimony from the original PET bottles was leaching into the soap.

A Diversity of "Isms"

Ultimately, it is the agenda with which we approach the making of things that must be truly diverse. To concentrate on any single criterion creates instability in the larger context, and represents what we call an "ism," an extreme position disconnected from the overall structure. And we know from human history the havoc an ism can create—think of the consequences of fascism, racism, sexism, Nazism, or terrorism.

Consider two manifestos that have shaped industrial systems: Adam Smith's *Inquiry into the Nature and Causes of the Wealth of Nations* (1776), and *The Communist Manifesto* by Karl Marx and Friedrich Engels (1848). In the first manifesto—written when England was still trying to monopolize her colonies and published the same year as the Declaration of Independence—Smith discounts empire and argues for the value of free trade. He links a country's wealth and productivity with general improvement, claiming that "Every man working for his own selfish interest will be led by an invisible hand to promote the public good." Smith was a man whose beliefs and work centered on moral as well as economic forces. Thus, the invisible hand he imagined would regulate commercial standards and ward off injustice would have been working in a market full of "moral" people making individual choices—an ideal of the eighteenth century, not necessarily a reality of the twenty-first.

Unfair distribution of wealth and worker exploitation inspired Marx and Engels to write *The Communist Manifesto*, in which they sounded an alarm for the need to address human rights and share economic wealth. "Masses of laborers, crowded into the factory, are organized like soldiers . . . they are daily and hourly enslaved by the machine, by the foreman, and, above all, by the individual bourgeois manufacturer himself." While capitalism had often ignored the interest of the worker in the pursuit of its economic goals, socialism, when single-mindedly pursued as an ism, also failed. If nothing belongs to anyone but the state, the individual can be diminished by the system. This happened in the former USSR, where government denied fundamental human rights such as freedom of speech. The environment also suffered: scientists have deemed 16 percent of the former Soviet state unsafe to inhabit, due to industrial pollution and contamination so severe it has been termed "ecocide."

In the United States, England, and other countries, capitalism flourished, in some places informed by an interest in social welfare combined with economic growth (for example, with Henry Ford's recognition that "cars cannot buy cars") and regulated to reduce pollution. But environmental problems grew. In 1962 Rachel Carson's *Silent Spring* promoted a new agenda—ecologism—that steadily gained adherents. Since then, in response to growing environmental concerns, individuals, communities, government agencies, and environmental groups have offered various strategies for protecting nature, conserving resources, and cleaning up pollution.

All three of these manifestos were inspired by a genuine desire to improve the human condition, and all three had their triumphs as well as their perceived failures. But taken to extremes—reduced to isms—the stances they inspired can neglect factors crucial to long-term success, such as social fairness, the diversity of human culture, the health of the environment. Carson sent an important warning to the world, but even ecological concern, stretched to an ism, can neglect social, cultural, and economic concerns to the detriment of the whole system.

"How can you work with *them*?" we are often asked, regarding our willingness to work with every sector of the economy, including big corporations. To which we sometimes reply, "How can you *not* work with them?" (We think of Emerson visiting Thoreau when he was jailed for not paying his taxes—part of his civil disobedience. "What are you doing in there?" Emerson is said to have asked, prompting Thoreau's famous retort: "What are you doing out there?")

Our questioners often believe that the interests of commerce and the environment are inherently in conflict, and that environmentalists who work with big businesses have sold out. And businesspeople have their own biases about environmentalists and social activists, whom they often see as extremists promoting ugly, troublesome, low-tech, and impossibly expensive designs and policies. The conventional wisdom seems to be that you sit on one side of the fence or the other.

Some philosophies marry two of the ostensibly competing

sectors, propounding the notion of a "social market economy," or "business for social responsibility," or "natural capitalism"—capitalism that takes into account the values of natural systems and resources, an idea famously associated with Herman Daly. Clearly these dyads can have a broadening effect. But too often they represent uneasy alliances, not true unions of purpose. Eco-effectiveness sees commerce as the engine of change, and honors its need to function quickly and productively. But it also recognizes that if commerce shuns environmental, social, and cultural concerns, it will produce a large-scale tragedy of the commons, destroying valuable natural and human resources for generations to come. Eco-effectiveness celebrates commerce *and* the commonweal in which it is rooted.

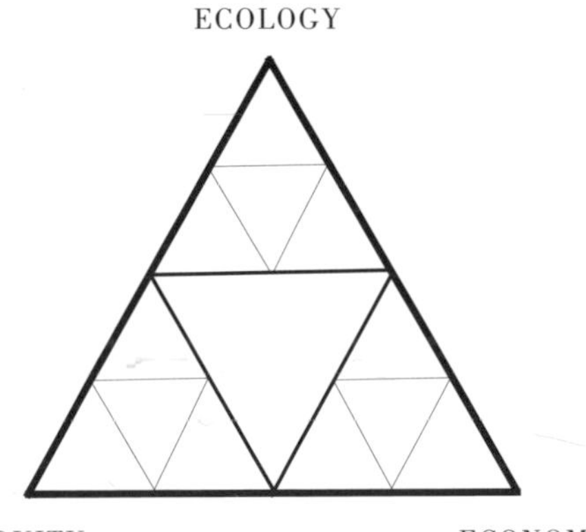

To make the process of engaging the various issues less abstract, we have created a visualization tool that allows us to conceptualize and creatively examine a proposed design's relationship to a multiplicity of factors, such as those we have been discussing in this chapter. It is based on a fractal tile, a form with no apparent scale that is composed of self-similar parts. This tool allows us to honor the questions raised by people in positions that lean dramatically toward one sector or another (Economy, for instance) as deserving of respect when taken in context. The fractal is a tool, not a symbol, and we have actively applied it to our own projects, ranging from the design of individual products, buildings, and factories to effects on whole towns, cities, even countries. As we plan a product or system, we move around the fractal, asking questions and looking for answers.

The extreme lower right represents what we would call the Economy/Economy sector. Here we are in the realm of an extremely pure capitalism, and the questions we ask would certainly include, Can I make or provide my product or service at a profit? We tell our commercial clients that if the answer is no, don't do it. As we see it, the role of commerce is to stay in business as it transforms. It is a commercial company's responsibility to provide shareholder value and increase wealth—but not at the expense of social structure and the natural world. We might go on to ask, How much do we have to pay people to get our product on the market and make a profit? If they are firmly entrenched in this corner—in the grip of an ism (pure capitalism)—they might consider moving production to a place where labor and transportation are as cheap as possible, and end the discussion there.

If they are committed to a more stable approach, however, we press on. We move over to the Economy/Equity sector, where we must consider questions of money and fairness; for instance, Are employees earning a living wage? (Here, again, sustainability is local: A living wage is going to be different wherever you live. From our perspective, it would be whatever it takes to raise a family.) Moving into the Equity/Economy sector, the emphasis shifts more toward fairness, so that we are seeing Economy through the lens of Equity, in a sense. Here we might ask, Are men and women being paid the same for the same work? In the extreme Equity corner, the questions are purely social—Are people treating one another with respect?—with no consideration of economics or ecology; this is where we can discuss issues such as racism or sexism.

Moving up to the Ecology corner of the Equity sector, the emphasis shifts again, with Equity still in the foreground, but Ecology is in the picture. Here the question might be, Is it fair to expose workers or customers to toxins in the workplace or in the products? Is it fair to have workers in offices where undefined materials are off-gassing, exposing them to potential health risks? We might also ask, How is this product going to affect future generations' health? Continuing into Ecology/Equity, we consider questions of ecosystem effects, not just in the workplace or at home, but with respect to the entire ecosystem: Is it fair to pollute a river or poison the air?

Now deep into the Ecology sector: Are we obeying nature's laws? Does waste equal food? Are we using current solar income? Are we sustaining not only our own species but all species? (The ism position in this corner would be Earth comes first, a tenet of "deep ecology"; do these things without worrying about Economy or Equity.) Then on around to Ecology/

Economy, where money reenters the frame: Is our ecological strategy economically fecund too? If we are designing a building that harnesses solar flows to make more energy than it needs to operate, the answer would be yes.

Finally, Economy/Ecology: this is where eco-efficiency is coming from, where we find people trying to be less bad, to do more with less while continuing to work within the existing economic paradigm. Still, as we have seen, eco-efficiency is a valuable tool in optimizing the broader eco-effective approach.

The Triple Top Line

The conventional design criteria are a tripod: cost, aesthetics, and performance. Can we profit from it? the company asks. Will the customer find it attractive? And will it work? Champions of "sustainable development" like to use a "triple bottom line" approach based on the tripod of Ecology, Equity, and Economy. This approach has had a major positive effect on efforts to incorporate sustainability concerns into corporate accountability. But in practice we find that it often appears to center only on economic considerations, with social or ecological benefits considered as an afterthought rather than given equal weight at the outset. Businesses calculate their conventional economic profitability and add to that what they perceive to be the social benefits, with, perhaps, some reduction in environmental damage—lower emissions, fewer materials sent to a landfill, reduced materials in the product itself. In other words, they assess their health as they always have—economically—and then tack on bonus points for eco-efficiency, reduced accidents or product liabilities, jobs created, and philanthropy.

If businesses are not using triple bottom line analysis as a strategic design tool, they are missing a rich opportunity. The real magic results when industry begins with all these questions, addressing them up front as "triple top line" questions rather than turning to them after the fact. Used as a design tool, the fractal allows the designer to create value in all three sectors. In fact, often a project that begins with pronounced concerns of Ecology or Equity (How do I create habitat? How can I create jobs?) can turn out to be tremendously productive financially in ways that would never have been imagined if you'd started from a purely economic perspective.

Nor are these criteria the only conceivable ones. High on our own lists is fun: Is the product a pleasure, not only to use but to discard? Once, in a conversation with Michael Dell, founder of Dell Computers, Bill observed that the elements we add to the basic business criteria of cost, performance, and aesthetics—ecological intelligence, justice, and fun—correspond to Thomas Jefferson's "life, liberty, and the pursuit of happiness." Yes, Dell responded, but noted we had left out a most important consideration: bandwidth.)

An Industrial Re-Evolution

Design that deeply respects diversity on all the levels we have discussed brings about a process of *industrial re-evolution*. Our products and processes can be most deeply effective when they are resonant with information and responses—when they most resemble the living world. Inventive machines that use the mechanisms of nature instead of harsh chemicals, concrete, or steel are a step in the right direction, but they are still *machines*—still a way of using technology (albeit benign technology) to harness nature to human purposes. The same could be said of our increasing use of cybertechnology, biotechnology, and nanotechnology to replace the functions of chemicals and brute force. The new technologies do not in themselves create industrial revolutions; unless we change their context, they are simply hyperefficient engines driving the steamship of the first Industrial Revolution to new extremes.

Even today, most cutting-edge environmental approaches are still based on the idea that human beings are inevitably destructive toward nature and must be curbed and contained. Even the idea of "natural capital" characterizes nature as a tool to be used for our benefit. This approach might have been valid two hundred years ago, when our species was developing its industrial systems, but now it cries out for rethinking. Otherwise, we are limited to efforts to slow the destruction of the natural world while we sustain the current industrial system of production and consumption for a few hundred years more. With human ingenuity and technological advances, we might even be able to create sustaining systems for our own species beyond that, after the natural world has greatly declined. But how exciting is sustainability? If a man characterized his relationship with his wife as sustainable, you might well pity them both.

Natural systems take from their environment, but they also give something back. The cherry tree drops its blossoms and leaves while it cycles water and makes oxygen; the ant community redistributes the nutrients throughout the soil. We can follow their cue to create a more inspiring engagement—a partnership—with nature. We can build factories whose products and by-products nourish the ecosystem with biodegradable material and recirculate technical materials instead of dumping, burning, or burying them. We can design systems that regulate themselves. Instead of using nature as a mere tool for human purposes, we can strive to become tools of nature who serve its agenda too. We can celebrate the fecundity in the world, instead of perpetuating a way of thinking and making that eliminates it. And there can be many of us and the things we make, because we have the right system—a creative, prosperous, intelligent, and fertile system—and, like the ants, we will be "effective."

Chapter Six

Putting Eco-Effectiveness into Practice

In May 1999, William Clay Ford, Jr., chairman of Ford Motor Company and great-grandson of its founder, Henry Ford, made a dramatic announcement: Ford's massive River Rouge factory in Dearborn, Michigan, an icon of the first Industrial Revolution, would undergo a $2 billion makeover to transform it into an icon of the next.

Henry Ford had bought the property when it was a marsh, and by the mid-1920s the plant began producing cars. In the following decades the River Rouge manufacturing plant grew to become one of the largest industrial complexes on the planet, fulfilling Ford's vision of a sprawling, vertically integrated facility capable of producing an automobile from start to finish. Coal, iron ore, rubber, and sand were brought in on barges from the Great Lakes. Blast furnaces, smelters, and rolling and stamping mills worked around the clock to produce the necessary materials. Working with Albert Kahn, his architect, Ford oversaw the design of powerhouses, body shops, assembly buildings, tool and die shops, an array of stockpiles, warehouses, factories, and associated infrastructure.

"The Rouge" was heralded as a marvel of manufacturing engineering and scale, and an emblem of modern industry. During the Depression, the factory even took on the job of taking apart used cars. A "disassembly line" was set up, with workers stripping each car of radiators, glass, tires, and upholstery as it moved down the line, until the steel body and chassis were dropped into an enormous baler. Admittedly the process was primitive and driven by brute force more than sophisticated design, but it was a striking illustration of "waste equals food" and an early step toward the reuse of industrial materials. Eventually the Rouge covered hundreds of acres and employed more than one hundred thousand people. It was a popular tourist destination and an inspiration to artists. In his photographs and paintings of the Rouge, Charles Sheeler portrayed the essence of a rational American manufacturing system. Painter Diego Rivera immortalized the factory from a worker's perspective in his astonishing murals installed at the Detroit Institute of the Arts.

By the end of the century, the facilities were showing their age. Although Ford's Mustang was still made there, the ranks of employees had dwindled to under seven thousand through divestiture, automation, and reduced integration. Over the years the plant's infrastructure had deteriorated. Its technology was outdated—the car plant, for example, was originally constructed in keeping with an assembly method in which parts were dropped down from floor to floor and assembled in a completed car on the bottom floor. Decades' worth of manufacturing processes had taken a toll on the soil and water. Major parts of the site had become brownfield—abandoned industrial land.

Ford Motor Company easily could have decided to do as their competitors had done—to close down the site, put a fence around it, and erect a new plant in a site where land was clean, cheap, and easily developed. Instead, it was committed to keeping a manufacturing operation going at the Rouge. In 1999 William Clay Ford, Jr., in his new post as chairman, took the commitment a step further. He looked at the rusting pipes and mounds of debris and took on the challenge (and the responsibility) of restoring it to a living environment. Instead of leaving the old mess and starting afresh somewhere else (moving on "like a pack of locusts," as one employee put it), Ford decided to help his company become native to its place.

Soon after becoming chairman, Ford had met with Bill to explore eco-effective thinking. A short meeting became an afternoon of exciting discussion, at the end of which Ford took Bill to his new office under construction on the twelfth floor, overlooking the Rouge in the distance. Did Bill think they could apply the principles they'd been discussing to that place—to go beyond recycling and "efficiency" to something truly new and inspiring? In May, Ford publicly asked Bill to lead the redesign of the River Rouge, from the ground up.

The first step was to create a "Rouge Room" in the basement of the company's headquarters, where the design team—which included representatives of all sectors of the company, along with outsiders like chemists, toxicologists, biologists, regulatory specialists, and union representatives—could come together. Their primary agenda was to create a set of goals, strategies, and ways of measuring progress, but they also just needed a setting that rendered visible their thinking process and encouraged them to raise the difficult questions. The walls were covered with working documents positioned under giant labels so that anyone walking through could see what was being considered in the way of socially, economically, and ecologically informed standards to measure the quality of air, habitat, community, energy use, employee relations, architecture, and, not least of all, production. Hundreds of employees came to the Rouge Room (jokingly referred to as a "peace room," as opposed to a "war room") during the process for structured meetings or simply to meet (often for other purposes) in a place suffused with so many of Ford's newly articulated intentions.

The company's commitment to financial security had been forged in the fire. Henry Ford had narrowly skirted bankruptcy during World War II, and seriously struggled to get the company back on its feet. Ever since then the bottom line has been a solid focus for everything the company does—every innovation must be good for profits. But the team had complete freedom to explore innovative ways of creating shareholder value, and the company's conventional decision-making process was to be informed by all aspects of the fractal tool we discussed in Chapter Five.

Once Bill Ford opened the door to the new thinking, hundreds of employees across all sectors of the company—in manufacturing, supply-chain management, purchasing, finance, design, environmental quality, regulatory compliance, and research and development (not only at River Rouge) began to come forward with ideas. There was internal resistance to overcome, to be sure, an entrenched skepticism that saw environmental strategies as at best extraneous to economics, and at worst as inherently uneconomic. One engineer burst into an early meeting saying, "I'm not here to talk to no eco-architect about no eco-architecture. I hear you want to put skylights all over the factory, and here at Ford we tar over skylights. And I hear you want to put grass on the roof. Now why am I here?" (He later turned out to be a hero of the project.) Also, as one scientific innovator within the company put it, the established scientific element at the company could be "like a fortress with a big moat." But, he added, "If there were no struggle around this, then by definition it would not have been very important."

Ford was already unique among automobile manufacturers in that, under then-director of environmental quality Tim O'Brien (and with Bill Ford's influence in his former role as a member of the environment committee), all of its plants had International Standards of Organization (ISO) environmental certifications that reflected their ability not only to monitor the quality of what they produced by standard metrics, but their environmental performance as well. The company had taken the additional step of requesting its suppliers to have the same. The ISO certification dictated that the company undertake a proactive investigation of environmental interests and concerns rather than relying on regulators to moderate it.

As Tim O'Brien himself pointed out, most manufacturers with old sites like the Rouge take a "don't ask, don't tell" approach, preferring not to examine their surroundings too carefully because any problem they discover will incur some obligation to act (and some vulnerability to lawsuits). When they do discover (or are forced to acknowledge) contamination, they usually remove the contaminated soil and bury it in a safe place, in compliance with EPA regulations. Such "scrape and bake" strategies may be efficient, but they are expensive and simply relocate the problems along with the topsoil.

Ford's design team said, "Let's assume the worst." When it found that there was indeed contamination at several of its plant sites, Ford negotiated with the government to experiment with treating its soil in a new way. It would remove and bury only the top layer of soil, then clean the deeper layers. It has been exploring innovative cleanup methods such as phytoremediation, a process that uses green plants to remove toxins from soil, and mycoremediation, or cleaning soil with mushrooms and fungi. From Rouge Room conception to implementation on the site, the approach is framed in positive, proactive terms—not "clean up" but "create healthy soil," for example. The phytoremediating plants are chosen for their indigenous as well as their toxin-cleansing properties. The health of the site is measured not in terms of meeting minimum government-imposed standards but with respect to things like the number of earthworms per cubic foot of soil, the diversity of birds and insects on the land and of aquatic species in a nearby river, and the attractiveness of the site to local residents. The work is governed by a compelling goal: creating a factory site where Ford employees' own children could safely play.

As the company looked at its new sustainability manufacturing agenda, it found more and more opportunities to improve environmental performance without conflicting with financial objectives, and these successes justified taking on more ambitious environmental challenges. Storm-water management and quality was a good start, because it is often taken for granted and appears to be inexpensive. But Ford discovered that storm-water management could be very expensive; regulations emerging from the Clean Water Act required new concrete pipes and treatment plants, threatening to cost the company up to $48 million. Instead, when the new plant is finished, it will have a green roof capable of holding two inches of rainwater, and porous parking lots that can also absorb and store water. Then the storm water will seep into a constructed marsh for purification by the plants, microbes, fungi, and other biota that live there. From the marsh the water will travel through swales—ditches full of native plants—on to the river, clear and clean. The storm water will take three days to seep to the river instead of heading there at once in a fierce, messy washout requiring quick, drastic measures. Instead of simply being a huge invisible liability, storm-water management is treated as a visible and enjoyable asset. The eco-effective approach cleans the water and the air, provides habitat, and enhances the beauty of the landscape while it saves the company a great deal of money—as much as $35 million by one estimation.

The redesign of the manufacturing facility embodies the company's commitment to social equity as well as to ecology and the economic bottom line. The old factory had become dark, dank, and unpleasant. Workers would keep one pair of shoes for use in the plant and one for street wear. In winter they might not see the sun for weeks, except on weekends. The company appreciates the fact that an enjoyable place to work is key to attracting a creative, diverse, and productive workforce. After visiting the Herman Miller factory Bill's architectural practice had designed in Michigan, the Ford team needed no more convincing: the new facility would be daylit—even the cafeteria, so that workers could get to daylight even on a short break—as Henry Ford's original factories had been, in an age of less energetic electrical systems. It would have high ceilings, plenty of unobstructed views, and (as a safety measure) supervisors' offices and team work rooms on a mezzanine to reduce the risk of accidents. The team also adopted Tom Kiser's way of viewing the building as a giant duct—and focusing on heating and cooling the people in the building rather than the building itself (see Chapter Five).

Ford sees River Rouge as a laboratory where it can test ideas it hopes will translate into a new way of designing for manufacture worldwide. Considering, for example, that the company alone owns approximately 200 million square feet of roofing around the world, successful innovations could be quickly implemented at industry-transforming scale. The specific solutions must grow out of and respond to local circumstances, however. A green roof might work in St. Petersburg, Florida, but not in St. Petersburg, Russia. Already the work at River Rouge has led to a review of other Ford plants where windmills and solar collectors could make economic sense if they are conceived as products of service within a total energy package. The company's overarching decision is to become native to each place. From that decision, local solutions follow, are adopted and adapted elsewhere as appropriate, and are continually revised and refined, effecting a profound process of change that may ultimately embrace every aspect of what a company makes and how it is produced, marketed, sold, and cycled on. A redesigned automobile factory may ultimately result in an entirely new notion of what an automobile is. It will take time to transform an industry so large, with such a complex infrastructure, but perhaps we will live to see a new automobile *disassembly* plant at the site of the first modern assembly plant.

Five Steps to Eco-Effectiveness

How does a company like Ford—with its long and distinguished history, its vast infrastructure, its large numbers of employees used to certain ways of doing things—begin to remake itself? It is not possible (nor would it necessarily be desirable) to simply sweep away long-established methods of working, designing, and decision-making. For the engineer who has always taken—indeed, has been trained his or her entire life to take—a traditional, linear, cradle-to-grave approach, focusing on one-size-fits-all tools and systems, and who expects to use materials and chemicals and energy as he or she has always done, the shift to new models and more diverse input can be unsettling. In the face of immediate deadlines and demands, such changes can seem messy, burdensome, and threatening, even overwhelming. But as Albert Einstein observed, if we are to solve the problems that plague us, our thinking must evolve beyond the level we were using when we created those problems in the first place.

Fortunately for human nature, in most cases change begins with a specific product, system, or problem and, driven by a commitment to putting eco-effective principles into action, grows incrementally. In our work, we have observed companies of all sizes, types, and cultures in this thrilling process of transition, and we have had ample opportunity to witness the steps they go through as they begin to retool their thinking and their actions in service to an eco-effective vision.

Step 1. Get "free of" known culprits.

Beginning to turn away from substances that are widely recognized as harmful is the step most individuals and industries take first as they move toward eco-effectiveness. We are so accustomed to hearing products touted as "phosphate free," "lead free," and "fragrance free" that the approach seems natural to us. Yet think how curious a practice it is. Imagine, for example, how your guests would react if, instead of describing the old family recipe you'd lovingly prepared, and the tasty ingredients you'd gone to such lengths to gather, you announced proudly that dinner would be "arsenic free."

It is important to acknowledge the potential absurdity of the approach and the less visible problems it may conceal. The detergent may be "free of" phosphates, but have they been replaced by something worse? The solvents that bind conventional printing inks are derived from problematic petrochemicals, but switching to a water base to make them "solvent free" may simply make it easier for the heavy metals that are still in the inks to enter the ecosystem. Bear in mind that positively selecting the ingredients of which a product is made, and how they are combined, is the goal.

Several years ago, we were asked to develop a chlorine-free container for a food company. When we thought about the project seriously, it became a bit of a sick joke, because we realized that simply being free of one thing did not necessarily make a product healthy and safe. As we have pointed out, the decision to make paper products that are chlorine-free means using virgin pulp rather than recycled paper, and even then, some naturally occurring chlorine will creep in. Moreover, the package contained other problematic substances—it had a polyurethane coating, for example, and there were heavy metals in the inks used to print on it—but these substances were not on anyone's well-publicized environmental hit list and so had yet to be perceived by the general public as dangerous. (We imagined the manufacturer could increase sales and save money and effort by simply announcing that the packaging was "plutonium free"!) Ironically, the manufacturer finally got its chlorine-free packaging only to discover chlorine-related dioxin in the food product itself.

Nevertheless, there are some substances that are known to be bioaccumulative and to cause such obvious harm that getting free of them is almost always a productive step. These are what we call X substances, and they include such materials as PVC, cadmium, lead, and mercury. Considering that the mercury in thermometers sold to hospitals and consumers in the United States each year is estimated to total 4.3 tons, and it takes only one gram to contaminate the fish in a twenty-acre lake, designing a mercury-free thermometer is a good thing. A well-publicized campaign is under way to eliminate mercury-based thermometers, but in fact that use accounts for only about 1 percent of the mercury used in the United States. By far the greatest amount is used for industrial switches of various kinds. A few auto manufacturers have phased out the use of mercury switches in cars—Volvo, which has been addressing these issues for years, also has a plan for phasing out PVC—but most have not. An industries-wide phaseout of mercury for this use is, from our perspective, crucial.

The decision to create products that are free of obviously harmful substances forms the rudiments of what we call a "design filter": a filter that is in the designer's head instead of on the ends of pipes. At this stage, the filter is fairly crude—equivalent to the decision not to include any items that might make your guests sick, or that they are known to be allergic to, when planning the menu for your dinner party. But it is a start.

Step 2. Follow informed personal preferences.

In the early 1980s, when Bill was designing the first of the so-called green offices for the Environmental Defense Fund's national headquarters, he sent questionnaires to manufacturers whose products he was considering using, asking them to explain exactly what the products contained. The questionnaires came back saying, in essence, "It's proprietary. It's legal. Go away." In the absence of data from the manufacturers themselves, Bill and his colleagues had to make choices based on their limited amount of information. For instance, they chose to tack down carpeting rather than to glue it, to avoid subjecting people to the various adhesives' unknown ingredients and effects. They would have preferred to use low-emission or no-emission adhesives that would allow the carpeting to be recycled, but those appeared not to exist. Likewise, they chose water-based paint. Their decision to use full-spectrum lighting meant importing bulbs from Germany, and while they preferred the quality of light (and knew it would make the workers feel good), they did not know much about the chemicals in the bulbs or the circumstances of their manufacture. For these and other design decisions, the team made choices based on the best information available to them, and on their aesthetic judgment. It would not do to select unattractive things just because they had more environmental authority—an ugly facility was not what they were hired to build.

When Bill began dealing with these issues as an architect in the 1970s and 1980s, he believed his job was to find the right things to put together, and he thought those things were already somewhere in the world. The problem was simply to find what and where they were. But it didn't take him long to discover that few truly eco-effective components for architecture and design existed, and he began to see that he could help to make them. By the time we met, Michael's thinking had evolved in a similar direction, and the future course of our work together was clear.

The truth is, we are standing in the middle of an enormous marketplace filled with ingredients that are largely undefined: we know little about what they are made of, and how. And based on what we do know, for the most part the news is not good; most of the products we have analyzed do not meet truly eco-effective design criteria. Yet decisions have to be made today, forcing upon the designer the difficult question of which materials are sound enough to use. People are coming for dinner in a few hours, and they expect to—need to—eat. Despite the astonishing paucity of healthy, nutritious ingredients, and the mystery surrounding, say, genetically modified crops (to carry the metaphor further), we cannot put off cooking until perfection has been achieved.

You might decide, as a personal preference, to be a vegetarian ("free of" meat), or not to consume meat from animals that have been fed hormones (another "free of" strategy). But what about the ingredients you do use? Being a vegetarian does not tell you exactly how the produce you are using has been grown or handled. You might prefer organically grown spinach to conventionally grown spinach, but without knowing more about the processor's packaging and transportation methods, you can't be certain that it is safer or better for the environment unless you grow it yourself. But we must begin somewhere, and odds are that as an initial step, considering these issues and expressing your preferences in the choices you make will result in greater eco-effectiveness than had you not considered them at all.

Many real-life decisions come down to comparing two things that are both less than ideal, as in the case of chlorine-free paper versus recycled paper. You may find yourself choosing between a petrochemical-based fabric and an "all natural" cotton that was produced with the help of large amounts of petrochemically generated nitrogen fertilizers and strip-mined radioactive phosphates, not to mention insecticides and herbicides. And beyond what you know lurk other troubling questions of social equity and broader ecological ramifications. When the choice is consistently between the frying pan and the fire, the chooser is apt to feel helpless and frustrated, which is why a more profound approach to redesign is critical. But in the meantime, there are ways to do the best with what we have, to make better choices.

Prefer ecological intelligence. Be as sure as you can that a product or substance does not contain or support substances and practices that are blatantly harmful to human and environmental health. When working on a building, for example, our architects might say that they prefer to use sustainably harvested wood. Without doing extensive research into individual sources that claim to supply such wood, they might decide to use a wood that comes with the Forest Stewardship Council seal of approval. We have not seen the particular forest where they are harvesting, and we don't know how deep their commitment to sustainability goes, but we have decided to go with the product based on what we know now, and the results will probably be better than had we not thought about the issue at all. And as Michael points out, a product that is, say, "free of PVC" or that in a general sense appears to have been made with care and consciousness points to a maker that has these issues as a mission.

In our work with an automobile maker, we've identified existing materials that are known to have some important positive qualities and are known not to have some common drawbacks: rubbers and new polymers and foam metals, "safer" metals such as magnesium, coatings and paints that won't put dioxin into the air. In general, we prefer products that can be taken back to the manufacturer and disassembled for reuse in technical production or, at the very least, returned to the industrial metabolism at a lower level—that is, "downcycled." We tend to opt for chemical products with fewer additives, especially stabilizers, antioxidants, antibacterial substances, and other "cleaning" solutions that are added to everything from cosmetics to paints to create the illusion of clean and healthy products. In truth, only a surgeon needs such protection; otherwise, these ingredients are only training microorganisms to become stronger while they exert unknown effects on ecological and human health. In general, because so few things seem to have been designed for indoor use, we try to choose ingredients that will minimize the risk of making people ill—that off-gas less, for example.

Prefer respect. The issue of respect is at the heart of eco-effective design, and although it is a difficult quality to quantify, it is manifested on a number of different levels, some of which may be readily apparent to the designer in search of material: respect for those who make the product, for the communities near where it is made, for those who handle and transport it, and ultimately for the customer.

This last is a complicated matter, because people's reasons for making choices in the marketplace—even so-called environmental choices—are not rational, and can easily be manipulated. Michael knows this firsthand, from a study he performed for Wella Industries, an international hair-care and cosmetic-products manufacturer that was trying to determine how people might be encouraged—through marketing and packaging—to choose environmentally friendly packaging for body lotions. A small but significant number of consumers chose to buy the lotion in a highly unattractive "eco" package shelved next to the identical product in its regular package, but the number who chose the "eco" package skyrocketed when it was placed next to an over-the-top "luxury" package for the very same product. People like the idea of buying something that makes them feel special and smart, and they recoil from products that make them feel crass and unintelligent. These complex motivations give manufacturers power to use for good and for ill. We are wise to beware of our own motivations when choosing materials, and we also can look for materials whose "advertising" matches their insides, again as indicative of a broader commitment to the issues that concern us.

Prefer delight, celebration, and fun. Another element we can attempt to assess—and perhaps the most readily apparent—is pleasure or delight. It's very important for ecologically intelligent products to be at the forefront of human expression. They can express the best of design creativity, adding pleasure and delight to life. Certainly they can accomplish more than simply making the customer feel guilty or bad in some way while immediate decisions are being made.

Step 3. Creating a "passive positive" list.

This is the point at which design begins to become truly eco-effective. Going beyond existing, readily available information as to the contents of a given product, we conduct a detailed inventory of the entire palette of materials used in a given product, and the substances it may give off in the course of its manufacture and use. What, if any, are their problematic or potentially problematic characteristics? Are they toxic? Carcinogenic? How is the product used, and what is its end state? What are the effects and possible effects on the local and global communities?

Once screened, substances are placed on the following lists in a kind of technical triage that assigns greater and less urgency to problematic substances:

The X list. As mentioned earlier, X-list substances include the most problematic ones—those that are teratogenic, mutagenic, carcinogenic, or otherwise harmful in direct and obvious ways to human and ecological health. It also includes substances strongly suspected to be harmful in these ways, even if they have not absolutely been proved to be. Certainly it should include the materials placed on the list of suspected carcinogens and other problematic substances (asbestos, benzene, vinyl chloride, antimony trioxide, chromium, and so forth) assembled by the International Agency for Research on Cancer (IARC) and Germany's Maximum Workplace Concentration (MAK) list. Substances placed on the X list are considered highest priorities for a complete phaseout and, if necessary and possible, replacement.

The gray list. The gray list contains problematic substances that are not quite so urgently in need of phaseout. The list also includes problematic substances that are essential for manufacture, and for which we currently have no viable substitutes. Cadmium, for example, is highly toxic, but for the time being, it continues to be used in the production of photovoltaic solar collectors. If these are made and marketed as products of service, with the manufacturer retaining ownership of the cadmium molecules as a technical nutrient, we might even consider this an appropriate, safe use of the material—at least until we can rethink the design of solar collectors in a more profound way. On the other hand, cadmium in the context of household batteries—which may end up in a garbage dump or, worse, airborne by a "waste-to-energy" incinerator—is a more urgently problematic use.

The P list. This is our "positive list," sometimes referred to as our "preferred list." It includes substances *actively defined* as healthy and safe for use. In general, we consider:

- acute oral or inhalative toxicity
- chronic toxicity
- whether the substance is a strong sensitizer
- whether the substance is a known or suspected carcinogen, mutagen, teratogen, or endocrine disruptor
- whether the substance is known or suspected to be bioaccumulative
- toxicity to water organisms (fish, daphnia, algae, bacteria) or soil organisms
- biodegradability
- potential for ozone-layer depletion
- whether all by-products meet the same criteria

For the moment, passive redesign of the product stays within its current framework of production; we are simply analyzing our ingredients and making substitutions where possible, aiming to select as many ingredients in the product as possible from the P list. We are rethinking what the product is made of, not what it fundamentally *is*—or how it is marketed and used. If you were planning dinner, you might be planning to not only use organically raised, hormone-free beef, but—having found spinach at a local farmer's market—to use the greens as well, and to eliminate the nuts you had planned to put in the cake because you've been alerted that one of your guests is allergic to them. But the menu would stay essentially the same.

For example, a manufacturer of polyester fabric, having discovered that the blue dye it is using is mutagenic and carcinogenic, might choose another, safer blue dye. We improve the existing product in increments, changing what we can without fundamentally reconceiving the product. In looking at a car, we might help (as we have) a manufacturer switch to upholstery and carpeting that are antimony-free, but we are not yet rethinking the fundamental design of the car. We might substitute a yellow paint without chromium for a yellow with chromium. We might omit a number of problematic, suspect, or simply unknown substances if we can make the product without them. We look as widely and deeply as we can at what *is*. Sometimes questionable substances in a product are not actually coming from the ingredients in the product but from something in or around the machinery used to make it, such as a machine lubricant, for which a less problematic substitute may be readily found.

Nevertheless, this step entails growing pains. Not yet having tackled a wholesale redesign of the product, the company has to match the quality of the old product while beginning to alter the ingredients list—the customer wants a blue just like the old blue. Just confronting the complexity of a given product can be daunting—imagine discovering (as we did) that a simple, everyday product used widely in manufacturing has 138 known or suspected hazardous ingredients. Yet this stage is the beginning of real change, and the inventory process can galvanize creativity. It may stimulate the development of a new product line that will avoid the problems associated with the old product. As such, it represents a paradigm shift and leads directly to . . .

Step 4. Activate the positive list.

Here's where redesign begins in earnest, where we stop trying to be less bad and start figuring out how to be good. Now you set out with eco-effective principles, so that the product is designed from beginning to end to become food for either biological or technical metabolisms. In culinary terms, you're no longer substituting ingredients—you've thrown the recipe out the window and are starting from scratch, with a basketful of tasty, nutritious ingredients that you'd love to cook with, and that give you all sorts of mouthwatering ideas.

If we are working with an automobile manufacturer, at this point we have learned all that we can about the car as it is. We know what it has been made of, and how the materials were put together. Now we are choosing new materials for it with a thought to how they can enter biological and technical cycles safely and prosperously. We might be choosing materials for the brake pads and rubber for the tires that can abrade safely and become true products of consumption. We might be upholstering the seats in "edible" fabric. We might be using biodegradable paints that can be scraped off on substrates of steel, or polymers that don't require tinting at all. We might be designing the car for disassembly, so that the steel, plastic, and other technical nutrients can once again be available to industry. We might be encoding information about all of the ingredients in the materials themselves, in a kind of "upcycling passport" that can be read by scanners and used productively by future generations. (This concept could be applied to many sectors of design and manufacturing. A new building could be given an upcycling passport that identifies all the substances used in its construction and indicates which are viable for future nutrient use and in which cycle.)

These are vast improvements on the current paradigm of "car." It will not end up on a scrap heap. And yet . . . it is still a car. And the current system of more and more cars on widening berths of asphalt is not necessarily ideal for the world of abundance we envision. (Buckminster Fuller used to joke that if extraterrestrial beings came in for a landing on Earth, their impression from ten thousand feet up would probably be that it was inhabited by cars.) Individually, cars can be fun, but terrible traffic jams and a world covered in asphalt are not. And so, having perfected the car as car, as nearly as we can, we move to . . .

Step 5. Reinvent.

Now we are doing more than designing for biological and technical cycles. We are recasting the design assignment: not "design a car" but "design a 'nutrivehicle.'" Instead of aiming to create cars with minimal or zero negative emissions, imagine cars designed to release *positive* emissions and generate other nutritious effects on the environment. The car's engine is treated like a chemical plant modeled on natural systems. Everything the car emits is nutritious for nature or industries. As it burns fuel, the water vapor in its emissions could be captured, turned back into water, and made use of. (Currently the average car emits approximately four fifths of a gallon of water vapor into the air for every gallon of gas it burns.) Instead of making the catalytic converter as small as possible, we might develop the means to use nitrous oxide as a fertilizer and configure our car to make and store as much as possible while driving. Instead of releasing the carbon the car produces when burning gasoline as carbon dioxide, why not store it as carbon black in canisters that could be sold to rubber manufacturers? Using fluid mechanics, tires could be designed to attract and capture harmful particles, thus cleaning the air instead of further dirtying it. And, of course, after the end of its useful life, all the car's materials go back to the biological or technical cycle.

Push the design assignment further: "Design a new transportation infrastructure." In other words, don't just reinvent the recipe, rethink the menu.

Most transportation infrastructure sprawls and devours valuable natural habitat or land that could be used for housing and agriculture. (The amount of space devoted to roads in Europe is currently equal to the space used for housing, and the two compete with agriculture.) Conventional development also depletes quality of life, with traffic noise, exhaust, and ugliness. A nutrivehicle that doesn't emit foul exhaust opens the way to a new approach to highways. They could be covered over, providing new green space for housing, agriculture, or recreation. (This might require less effort than it appears to. In many places, roadways are among the little public space still flanked by fields of green.)

If there are three times as many cars in twenty years as there are today on the planet, of course, it won't matter very much if they are highly efficient ultralight cars made from advanced carbon fibers and get a hundred miles to a gallon, or are even nutrivehicles. The planet will be crawling with cars, and we will need other options. A more far-ranging assignment? "Design transportation."

Sound fanciful? Of course. But remember, the car itself was a fanciful notion in a world of horse and carriage.

This final step has no absolute end point, and the results may be an entirely different kind of product than the one you began to work on. But it will be an evolution of that product in the sense that it addresses the limitations you became aware of as you moved through the previous steps. Design is based on the attempt to fulfill human needs in an evolving technical and cultural context. We begin by applying the active positive list to existing things, then to things that are only beginning to be imagined, or have not yet been conceived. When we optimize, we open our imaginations to radically new possibilities. We ask: What is the customer's need, how is the culture evolving, and how can these purposes be met by appealing and different kinds of products or services?

Five Guiding Principles

Transformation to an eco-effective vision doesn't happen all at once, and it requires plenty of trial and error—and time, effort, money, and creativity expended in many directions. Athletic-wear manufacturer Nike is one company that is taking a number of eco-effective initiatives to explore new material and new scenarios of product use and reuse. One of the company's agendas is to tan leather without questionable toxins, so that it is no longer a monstrous hybrid and can be safely composted after use. Because leather tanning affects so many products—including cars, furniture, and clothing—such an initiative could transform not one but several industries. Nike is also testing a clean new rubber compound that will be a biological nutrient and could likewise have a revolutionary impact on many industrial sectors. At the same time, the company is exploring innovations at the retrieval stage, attempting not only to make technical and biological nutrients but to put in place systems for retrieving them. The process is necessarily gradual—during this transitional period of introducing its new shoes, Nike separates and grinds the uppers, outsole, and cushioning midsole, and then works with licensees to create surfaces for sports activities (a fairly high-level use, still, as these materials offer protection from the elements as well as shock absorption). The goal remains upcycling, adapted to diverse locations and cultures, but not every avenue of exploration will pan out. As Darcy Winslow, Nike's global director of women's footwear, points out, in medium- and high-tech industries innovation typically has a success rate of 10 to 15 percent. The company is initiating several pilot programs to begin understanding the complexity of a product take-back program, with the expectation that one or several of those may end up working in the future. Nike sells products in approximately 110 countries, so the programs must be designed to incorporate regional and cultural relevancy.

There are some things design innovators and business leaders can do to help steer the transition at every stage and improve the odds of success:

Signal your intention. Commit to a new paradigm, rather than to an incremental improvement of the old. For example, when a business leader says, "We are going to make a solar-powered product," that is a signal strong enough for everyone to understand the company's positive intentions, particularly since total and immediate change is difficult in a market dominated by the status quo. In this case, the intention is not to be slightly more efficient, to improve on the old model, but to change the framework itself.

Employees "down on the ground" need to have this vision in place at the top, especially as they encounter resistance within the company. Tim O'Brien, newly promoted to vice president of real estate for Ford, says: "I know where to get 'yes': the twelfth floor," referring to the location of Ford's forward-thinking senior management team. "There may be argument on what the next steps will be at Ford, but there is no argument on the direction."

It is important, however, that signals of intention be founded on healthy principles, so that a company is sending signals not only about the transformation of physical materials but also about the transformation of values. For example, if the solar collectors powering a new solar-powered company are made with toxic heavy metals and no thought is given to their further use or disposal, then a materials problem has simply been substituted for an energy problem.

Restore. Strive for "good growth," not just economic growth. Think of the ideas we have presented here—and of designs in general—as seeds. Such seeds can take all manner of cultural, material, and even spiritual forms. For instance, a dilapidated neighborhood can be planted with such seeds as a new transit system, innovative ways of providing services that are not linked to waste and sprawl, water purification, the increase of green space and the planting of trees for cleaner air and beauty, the restoration of old and crumbling buildings, the revitalization of storefronts and marketplaces. On a smaller scale, buildings can be restorative: like a tree, they can purify water and send it out into the landscape in a purer form, accrue solar income for their own operations, provide habitat (for instance, designers can make roofs and courtyards attractive to birds), and give back to the environment. And, of course, design products that are restorative, as biological and technical nutrients.

Be ready to innovate further. No matter how good your product is, remember that perfection of an existing product is not necessarily the best investment one can make. Remember the Erie Canal, which took four years to build and was heralded as the height of efficiency in its day. What its builders and investors had not reckoned on was that the advent of cheap coal and steel would assure the canal's instant demise. The railroad was exponentially quicker, cheaper, and more convenient. By the time the canal was finished, the new niche and fitting-est technology for transportation had been developed.

When the fuel cell is becoming the automotive engine of choice in the automotive industry, those companies focused on increasing the performance and efficiency of the internal combustion engine might find themselves left behind. Is it time to keep making what you are making? Or is it time to create a new niche? Innovation requires noticing signals outside the company itself: signals in the community, the environment, and the world at large. Be open to "feedforward," not just feedback.

Understand and prepare for the learning curve. Recognize that change is difficult, messy, and takes extra materials and time. A good analogy is that of developing a wing. If you want to fly, at some point you need the sloppiness of additional materials, the redundancy—and a stretch for the research and development—to grow a wing. (Many scientists believe that wings evolved as a secondary use for limbs with feathers for warmth.) Biologist Stephen Jay Gould has captured this concept nicely in a way that can be useful to industry: "All biological structures (at all scales from genes to organs) maintain a capacity for massive redundancy—that is, for building more stuff or information than minimally needed to maintain an adaptation. The 'extra' material then becomes available for constructing evolutionary novelties because enough remains to perform the original, and still necessary, function." Form follows evolution.

You may not even know today what it is that you need to grow in the future, but if all of your resources are tied up in basic operations, there won't be anything extra to allow for innovation and experimentation. The ability to adapt and innovate requires a "loose fit"—room for growing in a new way. Rather than spend all its time and money fine-tuning an existing vehicle, for example, an automobile manufacturer might also be designing another car on the side: an innovative vehicle based on "feedforward." Innovative design takes time to evolve, but rest assured, in ten years the "perfect" vehicle of today will be a thing of the past, and if you don't have the new new thing, one of your competitors will.

Exert intergenerational responsibility. In 1789 Thomas Jefferson wrote a letter to James Madison in which he argued that a federal bond should be repaid within one generation of the debt, because, as he put it, "The earth belongs . . . to the living . . . No man can by natural right oblige the lands he occupied, or the persons who succeeded him in that occupation, to the payment of debts contracted by him. For if he could, he might, during his own life, eat up the usufruct of the lands for several generations to come, and then the lands would belong to the dead, and not to the living."

The context is different, but the logic is beautiful and timeless. Ask: How can we support and perpetuate the rights of all living things to share in a world of abundance? How can we love the children of all species—not just our own—for all time? Imagine what a world of prosperity and health in the future will look like, and begin designing for it right now. What would it mean to become, once again, native to this place, the Earth—the home of *all* our relations? This is going to take us all, and it is going to take forever. But then, that's the point.

Notes

Chapter One. A Question of Design

20 **"Citys . . . are nothing":** John Clare (1793–1864), "Letter to Messrs Taylor and Hessey, II," in *The Oxford Authors: John Clare*, edited by Eric Robinson and David Powell (Oxford and New York: Oxford University Press, 1984), 457.

21 **Consider cars:** James P. Womack, Daniel Jones, and Daniel Roos, *The Machine That Changed the World* (New York: Macmillan, 1990), 21–25.

22 **Henry Ford:** Quoted in Ray Batchelor, *Henry Ford: Mass Production, Modernism, and Design* (Manchester and New York: Manchester University Press, 1994), 20.

24 **"power, accuracy, economy":** Ibid., 41.

25 **"essences unchanged":** Ralph Waldo Emerson, "Nature," in *Selections from Ralph Waldo Emerson*, edited by Stephen E. Whicher (Boston: Houghton Mifflin, 1957), 22.

27 **more than 90 percent:** Robert Ayres and A. V. Neese, "Externalities: Economics and Thermodynamics," in *Economy and Ecology: Towards Sustainable Development*, edited by F. Archibugi and P. Nijkamp (Netherlands: Kluwer Academic, 1989), 93.

30 **mutations and infertility:** Marla Cone, "River Pollution Study Finds Hormonal Defects in Fish Science: Discovery in Britain Suggests Sewage Plants Worldwide May Cause Similar Reproductive-Tract Damage," *Los Angeles Times*, September 22, 1998.

31 **The reality of global warming:** DuPont, BP, Royal Dutch Shell, Ford, Daimler Chrysler, Texaco, and General Motors have withdrawn from the Global Climate Coalition, a group backed by industrialists that discounts global warming.

32 **Regulations for airborne pollutants:** The EPA is also adding statutes that manufacturers upwind of polluted areas are affected by regulations in those areas. See Matthew Wald, "Court Backs Most EPA Action in Pol-

luters in Central States," *The New York Times*, May 16, 2001, and Linda Greenhouse, "EPA's Authority on Air Rules Wins Supreme Court's Backing," *The New York Times*, February 8, 2001.

33 **asphalt and concrete:** In 1996 the impervious surfaces of the tristate metropolitan region around New York—the roads, buildings, parking lots, and nonliving parts—were measured at 30 percent. A generation ago this figure was 19 percent. The projection for 2020 is 45 percent. See Tony Hiss and Robert D. Yaro, *A Region at Risk: The Third Regional Plan for the New York–New Jersey–Connecticut Metropolitan Area* (Washington, D.C.: Island Press, 1996), 7.

35 **single-minded cultivation:** Wes Jackson has pointed out that the prairie as it was, with all of its diversity and grasses, actually produced more carbohydrates and protein per hectare than modern agriculture. But conventional agriculture has not engaged this rich ecosystem on its own terms.

35 **"a simplifier of ecosystems":** Paul R. Ehrlich, Anne H. Ehrlich, and John P. Holdren, *Ecoscience: Population, Resources, Environment* (San Francisco: W. H. Freeman, 1970), 628.

35 **returning complexity:** Many forms of "organic" agriculture that celebrate complexity and productivity are being developed around the world with rotations of animals and plants. For details, see the work of Sir Albert Howard, J. I. Rodale, Masanobu Fukuoka, Joel Salatin, and Michael Pollan. Another example of "homeostatic" (not single-purpose monocultural) farming, according to Wes Jackson, is the Amish agricultural method.

37 **a simplistic economic figure:** For an in-depth discussion of the GDP's failures and a presentation of new measurements for progress, see Clifford Cobb, Ted Halsted, and Jonathan Rowe, "If the GDP Is Up, Why Is America Down?," *Atlantic Monthly*, October 1995, 59.

38 **Since 1987:** Michael Braungart et al., "Poor Design Practices—Gaseous Emissions from Complex Products," *Project Report* (Hamburg, Germany: Hamburger Umweltinstitut, 1997), 47.

39 **"a formal risk assessment":** Wayne R. Orr and John W. Roberts,

"Everyday Exposure to Toxic Pollutants," *Scientific American*, February 1998, 90.

40 **legislation establishing:** Legislation is just beginning in Sweden.

41 **a child's swim wings:** Braungart et al., "Poor Design Practices," 49.

41 **Consider endocrine disruptors:** See Rachel Carson, *Silent Spring* (1962; rpt. New York: Penguin Group, 1997), and Theo Colburn, Dianne Dumanoski, and John Peterson Myers, *Our Stolen Future*, for an in-depth look at the effects of synthetic chemicals on human and ecological health.

Chapter Two. Why Being "Less Bad" Is No Good

45 **"I have read":** Thomas Malthus, *Population: The First Essay* (1798) (Ann Arbor: University of Michigan Press, 1959), 3, 49.

46 **"in Wildness":** Henry David Thoreau, "Walking" (1863), in *Walden and Other Writings*, edited by William Howarth (New York: Random House, 1981), 613.

46 **"When I submit":** Quoted in Max Oelshaeger, *The Idea of Wilderness: From Prehistory to the Age of Ecology* (New Haven: Yale University Press, 1992), 217.

48 **"hundreds of millions":** Paul R. Ehrlich, *The Population Bomb* (New York: Ballantine Books, 1968), xi, 39.

48 **"Then the fuse":** Paul R. Ehrlich and Anne H. Ehrlich, *The Population Explosion* (New York: Simon & Schuster, 1984), 9, 11, 180–81.

49 **"If the present":** Quoted in Donella H. Meadows, Dennis L. Meadows, and Jorgan Sanders, *Beyond the Limits: Confronting Global Collapse, Envisioning a Sustainable Future* (Post Mills, VT: Chelsea Green, 1992), xviii.

49 **"Minimize the use":** Ibid., 214.

49 **"The idea of unlimited growth":** Fritz Schumacher, *Small Is Beautiful: Economics as if People Mattered* (1973; rpt. New York: Harper and Row, 1989), 31, 34, 35, 39.

50 **"The simple truth":** R. Lilienfield and W. Rathje, *Use Less Stuff: Envi-*

ronmental Solutions for Who We Really Are (New York: Ballantine Books, 1998), 26, 74.

51 **"What we thought was boundless":** Joan Magretta, "Growth Through Sustainability: An Interview with Monsanto's CEO, Robert B. Shapiro," *Harvard Business Review* (January–February 1997), 82.

51 **"You must get the most":** Quoted in Joseph J. Romm, *Lean and Clean Management: How to Boost Profits and Productivity by Reducing Pollution* (New York: Kodansha America, 1994), 21.

52 **"Industries and industrial operations":** World Commission on Environment and Development, *Our Common Future* (Oxford and New York: Oxford University Press, 1987), 213.

52 **"Within a decade":** Stephan Schmidheiney, "Eco-Efficiency and Sustainable Development," *Risk Management* 43:7 (1996), 51.

53 **more than $750 million:** 3M, "Pollution Prevention Pays," http://www.3m.com/about3m/environment/policies_about3P.jhtml.

53 **almost 70 percent:** Gary Lee, "The Three R's of Manufacturing: Recycle, Reuse, Reduce Waste," *Washington Post*, February 5, 1996, A3.

54 **a groundbreaking report:** Theo Colborn, Dianne Dumanoski, and John Peterson Myers, *Our Stolen Future* (New York: Penguin Group, 1997), xvi.

54 **new research on particulates:** Mary Beth Regan, "The Dustup Over Dust," *Business Week*, December 2, 1996, 119.

59 **two fundamental syndromes:** Jane Jacobs, *Systems of Survival: A Dialogue on the Moral Foundations of Commerce and Politics* (New York: Vintage Books, 1992).

65 **no independent value:** For an interesting discussion of the "value" of efficiency, see James Hillman, *Kinds of Power: A Guide to Its Intelligent Uses* (New York: Doubleday, 1995), 33–44.

Chapter Three. Eco-Effectiveness

76 **a manager's job:** Peter Drucker, *The Effective Executive* (New York: Harper Business, 1986).

79 **some species of ant:** Erich Hoyt, *The Earth Dwellers: Adventures in the Land of Ants* (New York: Simon & Schuster, 1996), 27, 19.

80 **nature's services:** Gretchen C. Daily, introduction to *Nature's Services: Societal Dependence on Natural Ecosystems*, edited by Gretchen C. Daily (Washington, D.C.: Island Press, 1997), 4.

84 **"Nature being known":** Quoted in Clive Ponting, *A Green History of the World: The Environment and the Collapse of Great Civilizations* (New York: Penguin Books, 1991), 148.

Chapter Four. Waste Equals Food

94 **Rome's imperialism:** Sir Albert Howard notes that the "main causes" of Rome's decline "appear to have been fourfold: the constant drain on the manhood of the country side by the legions, which culminated in the two long wars with Carthage; the operations of the Roman capitalist landlords; failure to work out a balanced agriculture between crops and live stock and to maintain the fertility of the soil, the employment of Slaves instead of free labourers." Albert Howard, *An Agricultural Testament* (London: Oxford University Press, 1940), 8.

94 **"The central story":** William Cronon, *Nature's Metropolis: Chicago and the Great West* (New York and London: W. W. Norton, 1991), xv, 19.

95 **For centuries in Egypt:** For more details on the Egyptians' sustainable usage of the Nile, see Donald Worster, "Thinking Like a River," in *Meeting the Expectations of the Land*, edited by Wes Jackson, Wendell Berry, and Bruce Colman (San Francisco: North Point Press, 1984), 58–59.

96 **the Chinese perfected a system:** Also see F. H. King, *Farmers of Forty Centuries: Or, Permanent Agriculture in China, Korea, and Japan* (London: Jonathan Cape, 1925).

100 **Great Stink of London:** Clive Ponting, *A Green History of the World: The Environment and the Collapse of Great Civilizations* (New York: Penguin Books, 1991), 355.

105 **Most packaging:** Kyra Butzel, "Packaging's Bad 'Wrap,' " *Ecological Critique and Objectives in Design* 3:3 (1994), 101.

112 **rent-a-solvent:** Michael first proposed this concept in 1986. It is important to note, however, that it is not yet optimized; none of the companies

that have adopted the concept so far have yet completely rematerialized the solvent as a technical nutrient.

Chapter Five. Respect Diversity

120 **Think again of the ants:** Erich Hoyt, *The Earth Dwellers* (New York: Simon & Schuster, 1996), 211–13.

121 **ten species of ant wren:** John Terborgh, *Diversity and the Tropical Rain Forest* (New York: Scientific American Library, 1992), 70–71.

121 **A tapestry is the metaphor:** William K. Stevens, "Lost Rivets and Threads, and Ecosystems Pulled Apart," *The New York Times*, July 4, 2000.

147 **"Every man working":** Adam Smith, "Restraints on Particular Imports," in *An Inquiry into the Nature and Causes of the Wealth of Nations* (New York: Random House, 1937), 423.

148 **"Masses of laborers":** Karl Marx and Friedrich Engels, *The Communist Manifesto* (1848; rpt. New York: Simon & Schuster, 1964), 70.

148 **"ecocide":** See Murray Feshbach and Alfred Friendly, Jr., *Ecocide in the U.S.S.R.: Health and Nature Under Siege* (New York: Basic Books, 1992).

151 **a fractal tile:** Our fractal diagram is modeled on the Sierpinski gasket, named for the Polish mathematician who discovered it in 1919.

153 **"triple bottom line":** For more on this concept, see the work of John Elkington at www.sustainability.com.

Chapter Six. Putting Eco-Effectiveness into Practice

157 **a "disassembly line":** Charles Sorenson, *My Forty Years with Ford* (New York: W. W. Norton, 1956), pp. 174–75.

185 "**All biological structures**": Stephen Jay Gould, "Creating the Creators," *Discover*, October 1996, pp. 43–54.

Acknowledgments

We can't begin to acknowledge all the people who contributed to the ideas in these pages, and who are fostering and engaging with us in this discussion and in this work in the world—we would end up with a list that was endlessly long, and we know we might miss someone. So—we are deeply grateful to you all.

We would like to take particular note of the people who made this book possible, who worked on it, and who were involved in its creation. We reserve a special thanks for Lisa Williams, for her editorial endeavors during the whole of this project. We would also like to thank others whose vision and creativity have contributed in various ways to the shape this book took: Janine James; Charlie Melcher; our agent, Melanie Jackson; our editor, Becky Saletan, at North Point Press—which has instigated a new paradigm for bookmaking—and Anne Johnson, for her helpful research.

We especially want to acknowledge and thank our families. So here's to Michelle, Drew, and Ava from Bill, and to Monika, Jonas, Nora, and Stella from Michael. We celebrate your many gifts to us.

Sam Varnhagen

WILLIAM MCDONOUGH is an architect and the founding principal of William McDonough + Partners, Architecture and Community Design, based in Charlottesville, Virginia. From 1994 to 1999 he served as dean of the school of architecture at the University of Virginia. In 1999 *Time* magazine recognized him as a "Hero for the Planet," stating that "his utopianism is grounded in a unified philosophy that—in demonstrable and practical ways—is changing the design of the world." In 1996, he received the Presidential Award for Sustainable Development, the highest environmental honor given by the United States.

MICHAEL BRAUNGART is a chemist and the founder of the Environmental Protection Encouragement Agency (EPEA) in Hamburg, Germany. Prior to starting EPEA, he was the director of the chemistry section for Greenpeace. Since 1984 he has been lecturing at universities, businesses, and institutions around the world on critical new concepts for ecological chemistry and materials flow management. Dr. Braungart is the recipient of numerous honors, awards, and fellowships from the Heinz Endowment, the W. Alton Jones Foundation, and other organizations.

In 1995 the authors created McDonough Braungart Design Chemistry, a product and systems development firm assisting client companies in implementing their unique sustaining design protocol. Their clients include Ford Motor Company, Nike, Herman Miller, BASF, DesignTex, Pendleton, Volvo, and the city of Chicago. The company's Web site can be found at www.mbdc.com.